THE HISTORY OF THE 50TH

(Or THE QUEEN'S OWN) REGIMENT.

THE HISTORY OF THE 50TH

OR (THE QUEEN'S OWN) REGIMENT

From the earliest date to the Year 1881

BY

COLONEL FYLER

LATE COMMANDING THE REGIMENT

WITH COLOURED ILLUSTRATIONS, MAPS, AND PLANS,

AND

AN ILLUSTRATION OF THE OFFICERS' BREAST-PLATE, BEARING THE WORDS "QUO FATA VOCANT."

To

THE QUEEN'S

MOST EXCELLENT MAJESTY.

MADAM,

I feel that, while it would be impossible to add to the loyal devotion of "The Queen's Own" Regiment for your Most Gracious Majesty, the last mark of your Royal favour in accepting the Dedication of this Work, cannot fail to add to the *esprit de corps*, for which that Regiment has ever been distinguished.

To adopt the words addressed to Sir Charles Napier in 1852: It will "teach the young 50th, what he taught the old 50th; to fight to the death, for their beloved Queen, their country, and her people."

I have the honour to be

Your Majesty's
Most faithful subject and dutiful servant,

ARTHUR EVELYN FYLER,
Colonel, Retired Full Pay,
Late Commanding 50th (The Queen's Own)
Royal West Kent Regiment.

March, 1895.

PREFACE.

In compiling this work, I have availed myself of original despatches, and other documents at the Record Office, of the Orderly Room Records of the regiment, and of some valuable papers at the War Office, which I have kindly been permitted to make extracts from, and of some interesting sketches of uniforms which I have been allowed to reproduce. I am also much indebted to Mr. Milne, of Calverley, for much valuable information about old uniforms and colours.

In addition I am indebted to the following publications, from many of which I have made extracts :

For the siege of Ontario, I have obtained most of my information from a pamphlet on the conduct of Major-General Shirley, by William Alexander, published in 1758, and I have also made extracts from "Montcalm and Wolfe."

For the Seven Years' War, I am principally indebted to the account of the progress of the war given in the London Journals of the time, also to "Operations of the Allied Army," &c., &c., to the "History of the Grenadier Guards," by General Hamilton, and to Entick's and Lord Mahon's histories.

I have largely availed myself of "Wilson's Expedition," published in 1803, for the campaign in Egypt of 1801.

Sir William Napier's "The War in the Peninsula" has been my standard authority for that war, and I have taken the maps from his work; I have also been assisted by "The Adventures of Captain Patterson's 50th Regiment from 1807 to 1821," published in 1837, and by interesting papers at the War Office.

The campaigns in India have principally been compiled from original despatches, but I have derived great assistance from the daily journal kept by Lieutenant (now Colonel) Bellars, who was acting adjutant of the regiment during the greater part of the Sutlej campaign.

I have made extracts from Kinglake's "Invasion of the Crimea," especially with reference to the attack of a company of the 50th on the enemy's artillery train retreating after Inkermann. A certain amount of this campaign has been written from personal knowledge, assisted by Major-General Lock, and I am indebted to Colonel M. A. Clarke for his experience as a prisoner of war in Russia, which will be found in a paper in the Appendix.

I have studied the "War in New Zealand," by Fox, and this campaign has been corrected by Major Barker, who was adjutant through the war.

In the Appendix will be found short sketches of the life of Sir Charles Napier, for much of which I am indebted to his life by Sir W. Butler, and also of Sir Hudson Lowe, principally taken from "Napoleon at St. Helena," by W. Forsyth, M.A.

I have adopted the following abbreviations :—

O.R.R. for Orderly Room Records,

W.O.P. for War Office Paper, and

Kinglake for "Invasion of the Crimea," by Kinglake.

ERRATA.

The date at the top of pages 111 to 122 inclusive, and on the plan of Corunna, should be 1809 *instead of* 1808.

Page 123, Ensign Stewart should be among the list of killed.

Page 162, line 13, "See map, page 159" should be "See map, page 163."

Page 178, lines 18 and 19, "See map, page 173" should be "See upper map, page 171."

CONTENTS.

LIST OF ILLUSTRATIONS, MAPS, AND PLANS.

The Colour Plates In This Reprinting Are Placed After This Page

Private Soldier 50th REGT
about 1740.
From Sketch in War Office Book.

50th or 7th MARINES.
1741.
From Sketch in War Office Book.

1804 REGIMENTAL COLOUR.

1827 REGIMENTAL COLOUR.

INTRODUCTION.

From my first connection with the 50th Regiment, now more than forty years ago, I have heard repeated regrets that there were no historical records beyond that of the orderly room, which were very incomplete and not always available. On my retirement I determined to make an effort to remedy that deficiency, thinking that, as the last commanding officer of the old historic 50th, before it merged into the Royal West Kent, there was peculiar fitness in my becoming its historian, and I trust that the interest of the subject may atone for any defects.

I have heard many conversations among the officers about the glories of the past. There was a hazy idea of the regiment having been Marines, of a gallant charge at Vimiero, and a splendid service in India, where they earned the honourable soubriquet of "The Fighting Fiftieth"; but there was a vagueness and uncertainty about everything, only the best known facts standing prominently forward, like the brighter stars faintly visible in a misty sky, while many others scarcely less beautiful were completely obscured.

I had hardly calculated on the difficulties I had to encounter. Disraeli, in one of his works, makes a fearful monster the guardian of the threshold of the supernatural, but this "Dweller on the Threshold" had only to be resolutely faced to be made to disappear; so with the difficulties that beset a beginner, in searching among the rich mines of old historical literature in the reading-room of the British Museum, or of the original despatches and orders in the Record Office. The difficulties that appear insurmountable at first only require to be resolutely faced. The following may illustrate my meaning: at my first visit to the Record Office, after consulting the index, I asked for a certain correspondence, and was told that I could not see the documents between certain dates, without official permission from the department they belonged to. I at once applied unofficially to the War Office, and received a reply that they did not think permission was required, but would inquire. A week later I received a letter enclosing a message from the head of the Record Office, saying it was correct that these papers could not be shown without authority, but if I applied to him, he would give me permission to see them. Armed with this letter I again went to the Record Office, only to find that the writer of the letter was dead. However, this circumstance led to my introduction to the head of the Government search department, Mr. Charles Hall, to whom I have ever since been deeply indebted. From that time my difficulties were at an end.

Again, with reference to the British Museum, that gigantic collection of works, whose catalogues alone form an ordinary library, the perfect order in which everything is kept, and the great civility of the officials, reduces the labour of search to a minimum.

The plan I have adopted is to make each campaign in which the regiment was engaged a connected story *of that campaign*, that may, I hope, prove of more than regimental interest, and, while giving especial attention to every incident connected with the 50th, not to omit such mention of other regiments as may be necessary to complete the narrative.

I have endeavoured to trace the story of the 50th Regiment from its earliest date, until it ceased to bear that number in 1881; and, though it is impossible always to avoid dry details, I have, I hope, succeeded in covering the dry bones of history with interesting details of the gallant deeds in which the history of the 50th is peculiarly rich, as well as in bringing together much that is interesting, and some things that are new.

The following particulars are, I think, both new and interesting:

The pictures in the frontispiece of the old uniforms of the 50th Foot, and of the 50th or 7th Marines, page 1, *temp.* 1742, from an unpublished MS. book at the War Office.

The history of the siege of Ontario, by William Alexander, 1758, at which Shirley's 50th and Pepperell's 51st were taken prisoners, and both regiments afterwards disbanded.

An interesting account of the gallantry with which the 50th repeatedly charged and kept at bay a whole division of the French at the Pass of Lessessa, from a paper at the War Office which has never before been made public.

A portion of the 50th Regiment formed part of the 1st Battalion of the British Grenadiers in the Seven Years' War, and much interesting information is given about the Sutlej War, obtained from the diary and personal recollections of the acting adjutant of the regiment during the campaign.

The fact that fifteen volunteers from the French 70th Regiment enlisted in the 50th after Vimiero, and that the band of the 50th wore their long red plumes, as a trophy for some time afterwards, is highly interesting. A French colour pole and box was taken also in that action, and afterwards borne between the colours (Captain Patterson), but I have been unable to discover what became of it. It is curious to note, that the old colours were burnt at the head of the regiment, with military honours in 1804 and in 1815. (O. R. Records).

I have put a great deal of matter in the Appendix,

which, though of great interest, is irrelevant to the narrative, and might tend to make it tedious, especillay such dry details, as the stations occupied by the regiment at various times in Great Britain; the list of colonels and lieutenant-colonels (in which it may be noticed that Sir James Duffe commanded the regiment for over forty years, from 3rd August, 1798, to 23rd December, 1839); important official letters, &c.; also notices about Sir Charles Napier, Sir Hudson Lowe, Prince Waldemar's Cup, &c.

I was for some time under the impression, that the 50th Regiment might have been employed in the unfortunate expedition, that occupied Toulon in 1793;* the land forces being under the command of General O'Hara, who commanded the garrison of Gibraltar, at which place the 50th was quartered at the time, his A.D.C. being Captain Leith, 50th Regiment (afterwards Lieutenant-General Sir J. Leith, G.C.B.); and also from the fact of General Dundas (who commanded in Corsica) being second in command at Toulon. I thought this might have accounted for the palpable error in the Orderly Room Records, as to date of the regiment leaving Gibraltar, but I find that the 50th is not included in General Dundas's letter, claiming prize money for the Regiments engaged.

The siege of this place is memorable, as being the

* "The Broad Arrow" of November 24th, 1894 (p. 374), says the 50th Regiment "served at Toulon in 1793."

means of first bringing Napoleon Buonaparte to the front, and this was the only time, except at Waterloo, that he ever met British troops.

It is much to be regretted, that the old books and documents in possession of the regiment from its formation, were burnt at Naas Barracks on the 14th June, 1832, by authority of the Secretary of State for War, as these documents might have thrown some light on the subject.

Whatever difficulties I have had in compiling this history, will be more than repaid if it adds to the *esprit de corps* of the gallant 50th, by leading the young soldiers of the future, to study the glorious deeds of the past.

A. FYLER,
Colonel, Retired Full Pay.

Arlesey Bury,
Hitchen.

THE HISTORY OF THE 50TH (OR THE QUEEN'S OWN) REGIMENT.

CHAPTER I.

THE knowledge of the earliest date, on which a British corps bore the title of the 50th Regiment, would be most interesting. In early times, however, Regiments were designated by the names of their colonels only; and they were so styled even in official documents, up to the year 1753, and occasionally even later.

The first authentic record I can find is that of the 50th Foot, or the 7th Marines, which was raised on April 11th, 1741, under the command of Major-General Cornwall. Several regiments were officially the "50th" prior to 1741,* though the numbers were not used.

* In 1741, it is noted that the Regiments 54 to 59 were renumbered 43 to 48, and the seven new Regiments were numbered 49 to 55; this shows that a 50th Regiment of Foot must have existed prior to 1741, and a further closing up must have taken place later.

In the official Army List of 1740 the foot regiments, with the exception of the regiments of guards, are not numbered.

It contains 3 regiments of foot guards, 33 regiments of infantry, and 6 regiments of marines; 42 regiments in all.

Seven new regiments of infantry were ordered to be raised on the 11th April, 1741, the names of their colonels being given as Fowkes, Long, Houghton, Price, Mordaunt, Cholmondley, and De Graingues; and the same order directs four regiments of marines to be also raised, whose colonels were Major-General Cornwall, Colonels Hansom, Cawtell, and Jeffery. Thus, General Cornwall's would be the 50th Regiment of Infantry and the 7th Regiment of Marines.

An interesting plate is shown in the frontispiece taken from an unpublished book in the War Office, illustrating the uniform of that period (1740).

The uniform of the 50th being given, shows that that corps existed as a regiment of foot prior to April, 1741, when the 50th became marines, as there appears to have been no other 50th Regiment of Foot* until Shirley's regiment was raised in 1754.

After the peace of Aix la Chapelle, October,

* The Royal Artillery was numerically the 50th Regiment between these dates, but cannot be regarded as Foot.

1748, ten regiments were ordered to be disbanded, and Major-General Cornwall's regiment was included in this order, which was dated October 27th, 1748.

There seems, however, to have been some hesitation about disbanding this regiment, for the original order did not include it, and the order was eventually given in a separate letter, which implies that it was under consideration to retain the 50th Regiment alone as a regiment of marines. Short as its existence was, its services cannot have been unimportant, for it was employed as a regiment of marines, during a period of great naval activity, which included a war with France.

A War Office document of July, 1751, gives an official list of the regiments then in existence; all of which are numbered consecutively down to the 49th, after which comes the Royal Regiment of Artillery; and this record ends with the statement: "Forty-nine regiments of the line, and the Royal Regiment of Artillery; fifty regiments in all." At this period, therefore, the Royal Regiment of Artillery was the 50th Regiment. An order was signed by Mr. Fox on the 4th November, 1754 (see Appendix), and given to "Lieutenant-Colonel Ellison, of Her Majesty's 50th Regiment," ordering him to proceed to Virginia in a ship of war with his adjutant, and on arriving there to take the speediest method of going to Boston, in order to levy the regiment, of which William Shirley, Governor of New England, is appointed colonel.*

This regiment, though raised in America, and intended exclusively for service there, is styled Her Majesty's 50th Regiment in all official documents, and is included as such in the Army List of the day.

At this period (1755) a desultory war was being carried on between France and England, principally on the sea, and in North America, where the former had seized many important stations under British protection.

Colonel William Shirley was shortly afterwards promoted to major-general, and given the command of an expedition to Lake Erie and Ontario.

He was next appointed to the command of a force ordered to attack the French at Niagara, but this force never got further than Oswego; on Lake Ontario, where it was in imminent danger, until relieved by a successful attack of our Indian allies under Mr. Johnson. Major-General Shirley then abandoned any attempt to proceed further and returned to Albany, leaving the 50th and 51st Regiments, and 150 Jersey Provincials, under Colonel Mercer (51st Regiment), to entrench themselves for the winter at the embouchure of the Oswego river, where they suffered great privations from hunger

* The 51st Regiment, under Sir W. Pepperell, was raised in America at the same time.

disease, and exposure. Captain Vicars,* 50th Regiment, an officer invalided from Oswego, writing of this period, says: "He had passed the winter at Oswego, where he declared the dearth of food to have been such that several councils of war had been held on the question of abandoning the place from sheer starvation, and," he added, "had the poor fellows lived, they must have eaten one another. Some of the men were lodged in barracks, though without beds, while many lay all the winter in huts on the bare ground; scurvy and dysentery made frightful havoc. The garrison was so weak that the strongest guard we proposed to mount was a subaltern and 20 men, but we were seldom able to mount more than 16, and half these were obliged to have sticks in their hands to support them! The sentries were so weak that the men often fell down at their posts, and lay there till relieved. His company of 50 men was reduced to 10." (Montcalm and Wolf.)

In 1756 Major-General Shirley was superseded in the command-in-chief by the Earl of Loudoun, Major-General James Abercrombie being the second in command. In the summer of that year Colonel Braddock was sent with provisions to Oswego, and on his return, 29th July, 1756, he was unsuccessfully attacked, and learnt from one of the prisoners

* Captain John Vicars was second senior captain of Shirley's 50th in September, 1754.

taken that an attack was being organised on Oswego. A regiment was ordered to Oswego in support, but did not begin its march till the 12th August.

On the 4th August, 1756, the French, under Montcalm, began the embarkation of their troops to attack Oswego; and on the 10th, aided by the stranding of the British brigantine which should have guarded the coast, all were landed within half a league of Fort Ontario.* The French force consisted of about 1,800 regular troops, 2,500 Canadians, and 500 Indians, 32 pieces of cannon (12 to 24 pounders) and several large brass mortars.

About noon, on the 11th August, the enemy commenced their attack on Fort Ontario, garrisoned by 370 men of the 51st Regiment. About 3 p.m. on the 13th they had established their batteries within 60 yards of the Fort, and Colonel Mercer, considering it no longer tenable, destroyed the cannon, ammunition, and provisions, and retreated across the River Oswego, sending the garrison to reinforce

* An interesting and fuller account of this siege will be found in Appendix, taken from a work in the reading room of the British Museum containing three articles, one of them being "The Conduct of Major-General Shirley in North America, by William Alexander (called Lord Stirling), 1758." He presented his claim to the above title unsuccessfully in 1757, and was promoted to major-general in the American army in 1777, the year after the declaration of independence. The account of these operations is taken from the above work.

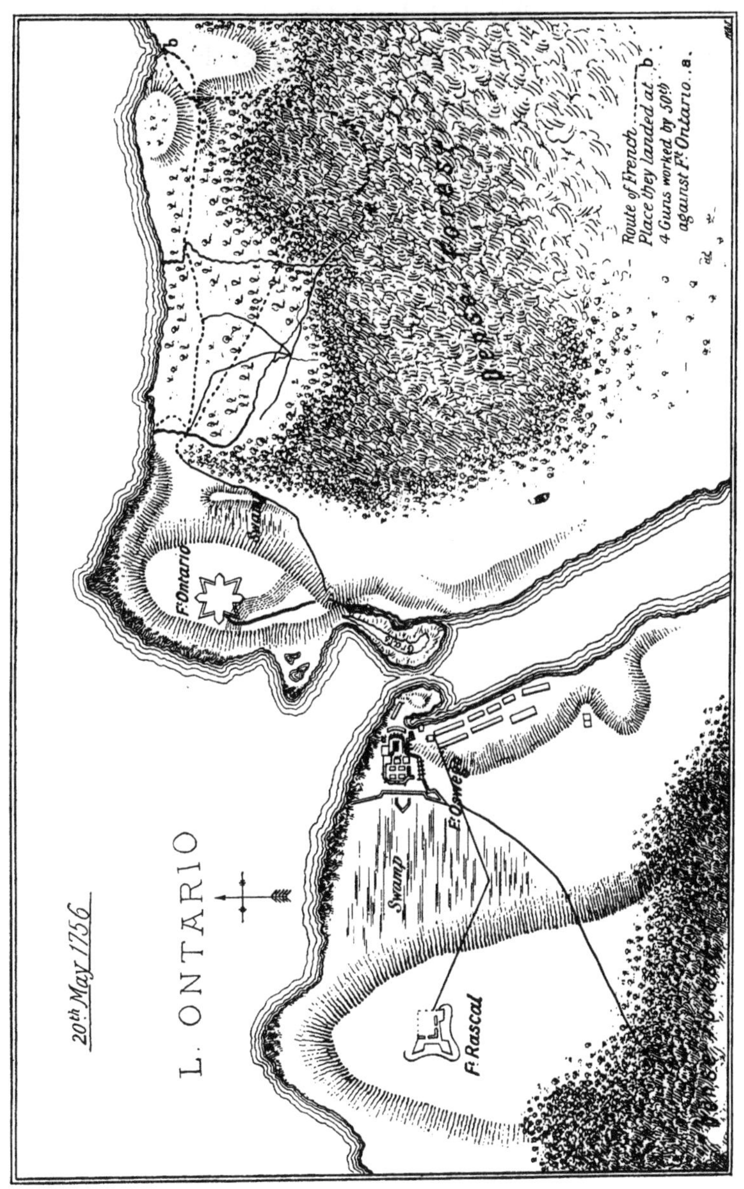
20th May 1756
L. ONTARIO
Ft Ontario
Swamp
Ft Oswego
Ft Rascal
Route of French
Place they landed at .. b
4 Guns worked by 50th
against Ft Ontario .. a

Fort Rascal. This left Fort Oswego garrisoned by the 50th Regiment, under the fire of Fort Ontario, on the opposite side of the river, by which it was commanded, and their batteries almost enfiladed; nevertheless the 50th appear to have worked their guns "remarkably well."

On the 14th the French crossed the river in three columns, by a ford, three-quarters of a mile higher up. Colonel Mercer was preparing to oppose the passage, when he was killed by a cannon shot.

Colonel Littlehales, who succeeded to the command, then called a council of war, which unanimously declared the works untenable. Two officers were then sent as envoys to the French commander to ask for terms. He replied—

"The English were an enemy he esteemed; that none but a brave nation would have thought of defending so weak a place so long, against such a strong train of artillery and superior numbers, and that the garrison might expect whatever terms were consistent with the service of his most Christian Majesty."

The garrison thereupon surrendered. "During the whole time of the siege, the soldiers behaved with a remarkable resolution and intrepidity against the enemy, exerting themselves to the utmost in the defence of the place in every part of duty; and it was with great reluctance, that they were persuaded by their officers to lay down their arms."

On the surrender of Oswego, on the 14th August, 1756, the greater part of the 50th (and 51st) Regiments became prisoners of war (War Office Letter, 25—1—1757, Appendix); part of them were sent to Canada, to be exchanged for French prisoners, and the remainder were transported to France. An order was therefore issued, to incorporate such parts of the non-commissioned officers and private men as were in North America (but not in Oswego), at the time, into other regiments of foot in North America; and to place the officers on half-pay, transferring such as were thought fit to other regiments, as vacancies occurred.

Such non-commissioned officers and privates of these two regiments, as were in England at the time, were at first ordered to be formed at Totness, into "bodies" of not less than 100 men, under 3 lieutenants and an ensign to each body; but, by a subsequent order, dated 2—2—1757, these men, amounting to 10 sergeants, 10 corporals, and 300 privates (War Office Letter, Appendix), were ordered to be "turned over to" the 2nd battalion of the 1st Royal Regiment of Foot, ordered to embark from Ireland, for the defence and protection of the colonies in North America.

And, finally, the War Office Letter already quoted, of 25—1—1757, provides that the establishments of the said 50th and 51st Regiments shall cease from the 25th December, 1756. It is curious to notice that the disbanding of the regiment is antedated a month; the officers, were, however, to receive full pay—until it was declared to them.*

The continued hostilities between France and England led to the struggle called the Seven Years' War, in consequence of which ten regiments of infantry were added to the regular army. The first of these regiments was numbered the 52nd Regiment, and Colonel James Abercrombie was appointed to the command, by a warrant from King George II., dated January 7th, 1756 (see Appendix). This regiment was first made up as follows:—

From the Foot Guards	10
From Lord Bertie's, the 7th Regiment .	110
From Lord Loudoun's, the 30th Regiment .	152

And the following recruits were raised:—

By the Earl of Sandwich	165
By the Earl of Fitzwilliam	58
By the Earl of Exeter	27
By the Duke of Ancaster	78
By the Regiment	140
Total	740

In consequence of the disbanding of the 50th Regiment, commanded by Major-General William Shirley, and of the 51st, commanded by Major-General Sir William Pepperel, the 52nd Regiment became the 50th.

* The soldiers of the 50th Regiment who were taken at Oswego, were either sent to Quebec to be exchanged, or embarked on board a French ship and taken to France, as prisoners of war.

I have been unable to find any official order for this change, though there is no doubt that it did take place, as the officers composing the 52nd, in the Army List of 1756, became those of the 50th in the Army Lists of 1757.

That the 52nd (or Abercrombie's) Regiment became the 50th is also proved by extracts from the London Gazettes of the time, in which the names of all the previous colonels are given. For instance, in the notification of the pension of a private soldier of the 50th Regiment, in the London Gazette of June 21st, 1760, we find the regiment described as "late Carr's, late Griffin's, late Hodgson's, late Abercrombie's." It is clear from the above that the date of the formation of the latest 50th Regiment, whose deeds are now historical, could not have been earlier than the 26th December, 1756, as Shirley's 50th were only disbanded the previous day. Before this date, Colonel James Abercrombie had been succeeded in the command of the 52nd, by Colonel Studholme Hodgson, who was the first colonel of the regiment,* which was then quartered at Ipswich.

The first notable incident in the career of the new regiment, occurred on the 5th of September,

* By an order of April 2nd, 1756, Major-General James Abercrombie was appointed to command in chief the land forces in America, until the arrival of the Earl of Loudoun. He afterwards commanded the 48th Regiment. Major-General Studholme Hodgson commanded the British force at the capture of Belle Isle, on the 7th June, 1761.

1757, on which date they embarked on board the fleet, under the command of Sir Edward Hawke. The expedition, which sailed under sealed orders, was intended for the capture of the French town and arsenal of Rochefort. It consisted of 16 ships of the line, with frigates and transports, and 10 regiments of foot under General Sir John Mordaunt. They sighted the isle of Oleran, which lies in front of Rochefort, on the 20th of September. On the 23rd, the islet of Aix was bombarded and taken, and on the 26th orders were given for the troops to hold themselves in readiness for a night attack on Rochefort,* but the night proved dark and stormy. The attack was countermanded, and, owing to the irresolution of the general in command, no further action was taken, and the fleet returned to England.

The 50th Regiment was landed at Spithead, and broken up into detachments about Guildford and Farnham.†

* The following is an extract from Entick's History :—

" Lieutenant-Colonel Sir William Boothby (who commanded the 50th on 5th September, 1764) took boat about twelve o'clock at night, though about two leagues from the shore and a great gale full against them, which would have endangered many of the boats, so crowded with men that they could only stand erect. They remained thus till about 2 p.m., when they were ordered to re-embark, as the other regiments could not be got ready." Sir William Boothby was at that date in the 30th Regiment, and Edward Carr and Studholme Hodgson, who both commanded the 50th, were originally in the 1st Foot Guards.

† The first record of the 50th Regiment being stationed at Maidstone is on the 7th of January, 1758.

The uniform worn by the 50th Regiment under Colonel Studholme Hodgson, appears to have been somewhat similar to that shown in the frontispiece (about 1740), and the black facings and white lace there shown, were worn till the year 1763, when the Army List shows the facings, to have been red with white linings and white lace—the 52nd Regiment being shown during this period, as wearing the old 50th uniform of black facings and white lace. Later lists continue to show this change until 1767, when both the 50th and 52nd Regiments, resume their former facings. In the course of this year, an order was issued directing the regimental number to be placed on all buttons; the button which was authorised for the 50th, contained the number in the centre surmounted by a crown, with single laurel leaves around the outer edge. In 1769, the lapels or loops on the coat became square at equal distances, with a red stripe down the centre.

About the end of the century, double Hanover loops were worn in pairs.

A sword breast-plate was introduced between 1770 and 1775, with an oval breast-plate—brass for private soldiers, and silver for officers.

The inspection return by General Dilkes, Dublin, 27th July, 1768, shows the dress of the officers of the 50th Regiment, to have consisted of a plain scarlet coat lapelled to the waist with black velvet, collar the same, small round black cuffs, silver buttons, silver epaulettes, white lining waistcoat and breeches, and silver laced hats. Very probably, this was the dress worn by officers in 1757, except that the waistcoat and breeches would have been red with white gaiters, which was the general costume until

SILVER BREAST-PLATE.

about 1765, and even later; after that period white waistcoat and breeches, with black gaiters, became general.

Trousers were used for field service in the Peninsula from about 1809, and at this period the officers of the first battalion, wore a silver breast-plate on the sword or cross-belt, having the motto "Quo fata vocant,"* a copy of which is shown in the accompanying engraving.

The second battalion wore the same breast-plate without the motto.

The engraving given on next page shows the brass breast-plate worn by private soldiers at the same period.

About this time also the crown on the silver button was surmounted by the lion, and both officers and men wore the short tunic or jacket—in the case of officers, double-breasted with two rows of silver buttons by pairs, and one button with a strip of black silk cord from it, on each side of the collar. Officers of the grenadier and light companies wore wings (instead of epaulettes) composed of silver curb chain edged with silver bullion fringe on black velvet, with gold grenades or bugles on the straps.

The light felt caps worn at this period, had a gilt plate in front, festooned across the front with crimson and gold cord, and a small red and white feather.

Private soldiers wore the same, but with plain white cords and tassels.

* I can find no authority for the adoption of this motto or for its abandonment, but the undoubted fact of its having been worn by the officers of the first battalion, during the Peninsula War, in which they greatly distinguished themselves, should form a strong claim for its restoration to the regiment.

I am indebted to Mr. S. M. Milne, for the loan of both these breast-plates, as well as for much valuable information about old uniforms.

Their red jackets were single-breasted, with ten loops of the white lace with a red stripe in the centre in pairs down the front, and on collar and shoulder-strap, slashes on cuffs and pocket, and a belt with brass breast-plate, as illustrated below.

BRASS BREAST-PLATE.

Sergeants had a similar coat of finer cloth, with white tape lace; and after 1802 silver chevrons on the right arm, with a belt and crimson sash, their arms being a sword and a long pike.

About the time of the accession of William IV. gold was substituted for silver. The officers' dress had recently been changed for a coatee, or scarlet tail-coat, on which captains and subalterns were for the first time, permitted to wear epaulettes on both shoulders. On becoming the "Queen's Own," blue velvet facings were adopted by the officers, of the colour known as Adelaide blue, and the soldiers changed their black facings for blue. A very handsome gold ornament was introduced, on the tail of each skirt of the officers' coatee, which consisted of "50" in gold on a blue ground, surrounded by a blue garter with "Honi soit," &c., and the edge of gold; a similar blue scroll underneath, with the words "Queen's Own" edged and lettered gold, surrounded by gold laurels and surmounted by a gold crown, all embroidered on red cloth.

The changes of uniform from this date having been much the same in all regiments, I need only mention further, the changes of the sword breast-plate of the officers.

About 1830, it was a square burnished gilt plate with the corners cut off, "50" in the centre in gold, on gold ground surrounded by a ring bearing the word "Regiment," and surmounted by a crown all in gold, while all round it, were the six ends of a silver star of the thistle order.*

Some years afterwards, this was changed for the

* A good specimen of this breast-plate, is in the possession of Mr. S. M. Milne.

breast-plate worn by officers in the Crimea, composed of a square gold bevilled plate, with "50" in the centre in matted silver, the star and centre and the sphynx underneath being bright silver, and the girdle round "50" gilt.*

* A specimen of this is in the Royal United Service Institution.

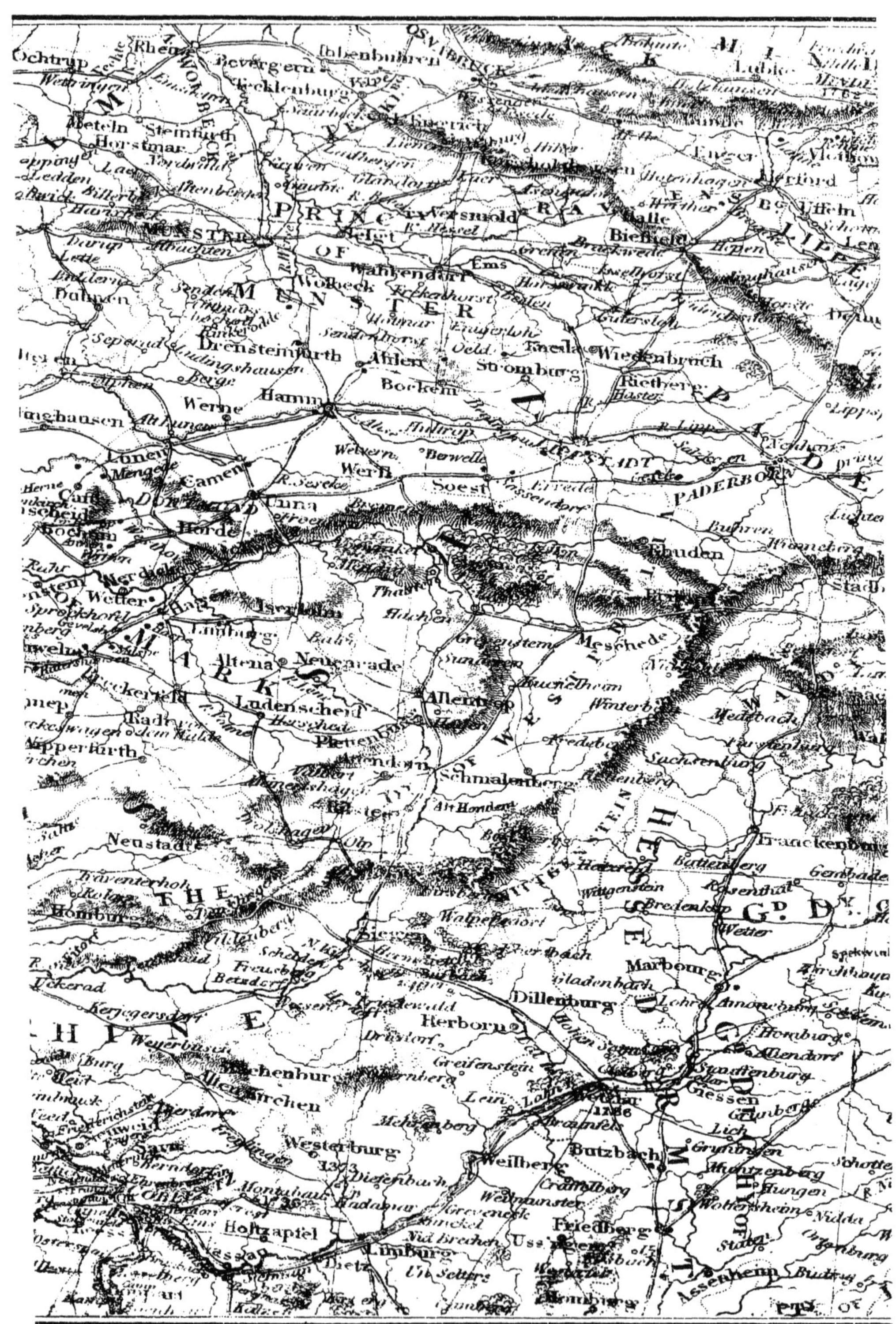

Ochtrup
Rheine
Bevergern
Ibbenbuhren
Tecklenburg
Beteln
Horstmar
Steinfurth
Enger
Herford
PRINC
OF
MUNSTER
Versmold
Halle
Bielefeld
Warendorf
Wolbeck
Ems
Dulmen
Drensteinfurth
Ahlen
Stromburg
Wiedenbruch
Rietberg
Bockem
Hamm
Werne
Camen
Unna
Werl
Soest
PADERBORN
Horde
Wetter
Iserlohn
Altena
Neuenrade
Meschede
MARK
Ludenscheid
Allendorp
Schmalenberg
Neustadt
Homburg
Siegen
Marbourg
Dillenburg
Herborn
Giessen
Butzbach
Weilberg
Friedberg
Westerburg
Holzapfel
Limburg
Franckenberg
Gᴰ Dʸ
Wetter

Map of Hessi

Vincent Brooks, Day & Son Photo

E-CASSEL.

CHAPTER II.

SEVEN YEARS' WAR—1760.

In the year 1757 France, by virtue of the Treaty of Closter Seven, was in undisputed possession of the Electorate of Hanover, then united to the British Crown. England was at that time in alliance with Prussia, whilst Russia and Austria opposed them. After the total defeat of the French at the battle of Rosbach, the Hanoverians rose against them and drove them out, and Pitt tardily consented to send British troops to aid the Hanoverian army, which was under the command of Ferdinand, Duke of Brunswick and Luneberg.

The 50th Regiment did not take part in the campaign, until after the Battle of Minden. They received orders to join the army in Germany on the 8th of May, 1760, arriving there in the following June, and joined the corps d'armée of Prince Ferdinand on the 20th of that month. The regiment speedily had an opportunity of distinguishing themselves; and a portion of them incorporated with the

1st Battalion of British Grenadiers were prominently engaged in the campaign.*

It consisted of parts of Hodgson's (5th), Barrington's (8th), Bockland's (11th), Cornwallis' (24th), Griffin's (33rd), and Carr's (50th). (London Gazette.)

The allied army, which the 50th Regiment had now joined, quitted their camp at Fritzlar on the Eder on the 24th of June, and on the 27th they took up a position between Treissa and Zigenhayn on the right of the Schwalm, having that river in their front. M. de Broglio, who commanded one of the French armies, opposed them with his right at Neustatt, and his left at Speckwinckel, the two armies being about eight miles apart.

On the 3rd of July M. de Broglio began a movement to his left, taking the small fortified places of Rosenthal and Frankenberg, which the allies met by

* The brigade was at first styled the Brigade of British Grenadiers and Highlanders, but it was afterwards called the British (and occasionally the English) Grenadiers.

During the years 1761 and 1762 they formed the first brigade of Lord Granby's corps under Colonel Beckwith.

Part of the 50th Regiment was incorporated with the first battalion, which was under Major Daulhart at the battle of Warburg, but afterwards under Major Welsh.

Probably it comprised the two flank companies of the 50th Regiment. The orderly room records of the 50th evidently refer to this corps when they say: "The flank companies were engaged in several affairs during the campaign."

It is unfortunate that in the lists of killed and wounded of this corps no referencs is made to the regiments to which they belonged, except at Warburg.

sending parties to secure Nieder-Urff and Wildungen on the 4th.

Prince Ferdinand, finding the French army of the Lower Rhine under M. de St. Germain was also advancing from Lunen on the Lippe river, viâ Dortmond, towards Brilon and Corbach, broke up his camp at 3 p.m. on the 8th July and retired, arriving on the 9th on the heights of Braunen (or Brauna), near Wildungen, where he encamped. The advanced corps, under the Hereditary Prince, (nephew * of Prince Ferdinand), reinforced by some extra battalions (which included the 50th Regiment) and some squadrons under Major-General Griffin,† was sent forward the same day to secure Saxenhausen.

M. de Broglio, judging that Prince Ferdinand's design in retreating was to secure the heights of Corbach, and knowing that such also was the object of the French army under M. de St. Germain, ordered the advanced guard of his army to push forward for that place, and followed them with the main body by forced marches.

At 2 o'clock on the morning of the 10th July the Hereditary Prince with the vanguard, including Major-General Griffin's force, marched to Corbach, but found that the advanced guard of M. de Broglio's army and the van of M. de St. Germain's had

* One authority says son of Prince Ferdinand.

† Major-General Carr had just succeeded Major-General Griffin in the command of the 50th Regiment.

arrived before them and taken possession of the heights. The Hereditary Prince, however, conceiving that the force opposed to him was not very strong, and estimating their whole force not to exceed 10,000 to 15,000 infantry and 17 squadrons, formed the idea of driving them back. With this view the advance corps made a dashing attack upon the enemy, which became extremely hot about 2 o'clock; but the enemy being continually reinforced, and having the advantage of numbers and position, the Prince found it impossible to dislodge them; and an order to retire having been received from Prince Ferdinand, whose main army had now arrived at Saxenhausen, the Hereditary Prince commenced a retreat, which proved very disastrous, for they were vigorously followed up by the now greatly superior forces of the enemy, and lost 12 cannon, 4 howitzers, 27 officers and 797 non-commissioned officers and men killed, wounded, and missing. The loss of the 50th was 2 privates killed, Lieutenant Cathcart and 7 privates wounded, and 6 privates missing.

The following extract from an official letter from Prince Ferdinand, dated the 11th July, 1760, shows that the 50th Regiment took a prominent part in the above engagement:—

"General Count de Kilmansegge greatly distinguished himself in this affair, as well as Major-General Griffin, with the two English battalions, Carr's (50th) and Brudenel's (51st)." (Translation.)

The French army encamped on the heights of Corbach, and that of Prince Ferdinand at Saxenhausen, their left not more than a mile and a half from the enemy, and separated from them by an almost impracticable ravine.

After some skirmishes the French army struck their tents on the 24th July, and moved towards the Dymel river.

Prince Ferdinand marched to Wolfshagen on the road to Cassel.

The French army now broke up into three columns, one of them under Prince Xavier threatening Cassel, while another endeavoured to cut off General Sporcken's command from the rest of the allied army.

The subjoined despatch from Prince Ferdinand, dated head-quarters at Warburg, 1st August, 1760, shows the further action of his column. (See map of Cassel and vicinity.)

The Chevalier de Muy having passed the Dymel at Stadbergen, with the reserve of the French army under his command (amounting as is supposed to upwards of 35,000 men), extending itself down to the river, in order to cut off our communications with Westphalia, while Marshal Broglio advanced with the main army towards our camp at Kalle, as Prince Xavier of Saxony did likewise towards Cassel, which place General Kilmanseggc was left with a body of troops to protect.

H.S.H. passed the Dymel in the night of the 30th between Liebenau and Dringelbourg.

The Hereditary Prince passed that river on the 29th to reinforce General Sporcken, and took post between Liebenau and Corbeke,* the whole force then consisting of 24 battalions and 22 squadrons.

On the 30th his Highness reconnoitred De Muy's camp between Warbourg and Ocksendorf.

About 5 the next morning 31st July, the grand army assembled and formed on the heights near Corbeke, while the Hereditary Prince was marching in two columns in order to turn the enemy's flank, which was attacked almost at the same time by the Hereditary Prince, and M. de Sporcken; and after a very sharp engagement the enemy was at last obliged by our continual fire to give way, and fall back upon Warbourg.

The army was marching in the meanwhile with the greatest expedition to attack the enemy in front, but as the infantry could not get up in time, the Marquis of Granby was ordered to advance with the cavalry of the right, and the English artillery, &c., &c.

The town of Warbourg was then attacked by the Legion Britannique, and the enemy finding themselves thus attacked on both their flanks, in front, and in rear, retired with the utmost precipitation, and with the loss of many men, both from the fire

* Places in vicinity of Cassel will be found in large map at p. 42.

of our artillery and from the charges of our cavalry. Many of them were drowned in attempting to ford the Dymel, &c.

Return of the killed, wounded, and missing of the British Grenadiers and Highlanders on the field of Warbourg, 31st July, 1760, from the London Gazette, 19th August, 1760 :—

INFANTRY.

1st Battalion, commanded by Major Daulhart.

Regiment.	Killed.	Wounded.	Missing.
Hodgson's 5th	4 Pts.	2 Of. 26 Pts.	
Barrington's 8th	1 N.C.O. 2 Pts.	1 Of. 13 Pts.	1 Pt.
Bocland's 11th	6 Pts.	21 Pts.	6 Pts.
Cornwallis' 24th	1 Of. 1 N.C.O. 3 Pts.	1 Of. 21 Pts.	3 Pts.
Griffin's 33rd	6 Pts.	33 Pts.	
Carr's 50th	4 Pts.	2 N.C.O. 14 Pts.	

Total killed, 1 Of., 2 N.C.O., 25 Pts. ; wounded, 4 Of., 2 N.C.O., 128 Pts.

2nd Battalion, commanded by Major Maxwell.

Regiment.	Killed.	Wounded.	Missing.
Napier's 12th	1 N.C.O. 15 Pts.	1 Of. 3 N.C.O. 32 Pts.	
Kingsley's 20th	2 N C.O. 13 Pts.	1 Of. 3 N.C.O. 35 Pts.	
Boscawen's 23rd	1 N.C.O. 11 Pts.	2 Of. 19 Pts.	
Horne's 25th*	1 N.C.O. 7 Pts.	1 Of. 1 N.C.O. 24 Pts.	4 Pts.
Stuart's 37th	10 Pts.	2 Of. 1 N.C.O. 19 Pts.	
Brudenel's 51st	9 Pts.	1 Of. 1 N.C.O. 22 Pts.	

Total killed, 5 N.C.O., 65 Pts. ; wounded, 1 Of., 9 N.C.O., 151 Pts.

* Afterwards Erskine's.

The enemy are reported to have left 1,500 on the field of battle, with a loss of an equal number of prisoners, 10 pieces of cannon, and some colours. The fate of the action was decided by an attack upon the enemy's left, in which the brigade of British Grenadiers and Highlanders took part. Prince Ferdinand, in a further despatch to His Britannic Majesty on this affair, says: "Colonel Beckwith, who commanded the brigade of British Grenadiers and Scotch Highlanders, distinguished himself greatly, and was wounded in the head."

The chief loss of the allies fell upon the British Grenadiers and Highlanders, who had 415 men killed and wounded out of a total of 590 men lost by the British; the casualties in the rest of the allied army being trifling. The loss to the 50th Regiment was 4 rank and file killed, 2 non-commissioned officers and fourteen rank and file wounded.*

The majority of the British infantry under General Waldegrave were unable to get up in time, although they "pressed their march as much as possible;" many of the men from the heat of the weather, and over-straining to get on through morassy and very difficult ground, dropped down dead on their march.

Prince Ferdinand's army then encamped on the heights of Warburg, with the Dymel in their front;

* The loss of the 50th in this action occurred in the 1st Battalion of British Grenadiers. (See list of killed and wounded of that corps).

the enemy being on the opposite side separated by the river.

The above success was, however, counterbalanced by the loss of the fortified town of Cassel, defended by General Kilmansegge with 12 battalions and 6 squadrons.

Prince Ferdinand, writing to Count Holdenesse (August 1) on the subject, says: "I am in despair, my lord, to announce to you the loss of the town of Cassel, taken by Prince Xavier at the moment that I sent the news of the defeat of M. de Muy."

The French army left their camp on the Dymel in the night between the 21st and 22nd of August. The Hereditary Prince crossed that river on the 22nd at the head of 12,000 men, and endeavoured to gain the left flank of the enemy. His advanced troops came up with their rear guard at Zierenberg, and after the light troops on each side had been engaged with indifferent success, the Hereditary Prince arrived in person, with the Greys and Inniskillings, supported by the English Grenadiers, and put an end to the affair in a quarter of an hour by forcing the enemy to a precipitate flight with great loss. The same evening 12 British regiments* and 10 squadrons passed the Dymel in pursuit, and encamped near Wilda.

Three second battalions of the Guards arrived at Weser on the 30th July, and overtook the British contingent at Buhne, near Warburg, on the 25th of

* The 50th Regiment must have been one of these battalions.

August, Buhne being at the time the head-quarters of Prince Ferdinand.

Lieut.-Gen. the Hon. Seymour Conway's corps was then arranged as under:—

1st Brig., Brig.-Gen. Cæsar.		2nd Brig., Brig.-Gen. Townshend.		
Grenadiers of the Guard.		8th Regiment,		Barrington.
2nd Battalion 1st Guards.		25th	,,	Erskine.
Do.	Coldstreams.	50th	,,	Carr.
Do.	3rd Guards.	20th	,,	Kingsley.

Cavalry Brigade, Brig.-Gen. Douglas.

Bland's (1st Dragoon Guards), 3 squadrons.

Howard's, 2 squadrons.

Waldegrave's (5th Dragoon Guards), 2 squadrons.

("History of the Grenadier Guards," by General Hamilton.)

Nothing further that affected the 50th Regiment occurred during this year. As far as I can ascertain only one battalion of British Grenadiers (Maxwell's) being employed at the surprise and gallant capture of the town of Zierenberg on the 5th September.*

* The French account of the surprise of Zierenberg gives the following:—"The column of English Grenadiers advanced in great order, with their bayonets fixed, without firing a musket, by the two streets that lead to the churchyard, the only square or open place in the whole town. These Grenadiers advanced to the churchyard in the greatest order and the greatest silence. The night was so dark that they formed at the side of the French troops, who for some time imagined them to be their own piquet, but at last, discovering their mistake, a fierce encounter ensued with bayonets, and several were killed or wounded. The English being superior in numbers drove the infantry from the churchyard."—*Operations of the Allied Army.*

The roads having become impracticable the troops went into winter quarters early in December—the Guards at Paddeborn, and the rest of the troops in that country and along the Dymel and Weser—the Marquis of Granby's head-quarters being at Corvey on the Weser.

The following year the 50th Regiment continued in the same brigade and division.*

* The following state of the 50th Regiment, from the official return of November 1, 1761, will be interesting:—

Present 50th (Carr's).

0	Colonel
0	Lieut.-Colonel
0	Major
6	Captains
15	Lieutenants
6	Ensigns
1	Chaplain
1	Adjutant
1	Quartermaster
0	Surgeon
2	Mates
33	Serjeants
13	Drummers

Privates.

477	Fit for duty
56	Sick present
133	Do. in hospital
8	On command
0	Recruiting
0	On furlough
674	Total.

Officers absent.	By whose leave.
Lieutenant-General Carr	Lord Granby's.
Lieutenant-Colonel Prescot	Lord Granby's.
Major Barry	Sick in hospital.
Captain Sir A. Hope	Sick in hospital.
Captain Warburton	On duty at Bremen.
Lieutenant Baskerville	Lord Granby's.
Lieutenant Ross	Quartermaster at Bremen.
Ensign Cook	Lord Granby's.
Ensign Laycraft	Not joined.

The early part of this year found the French in possession of Hesse, with several well-fortified places and numerous magazines. Their left was at Wesel* on the Lippe river, their right in the fortress of Gottingen, with strong posts in rear and on both flanks.

The danger of being surrounded to which Prince Ferdinand was thus exposed, induced him to try the effect of a winter campaign. He therefore drew together his troops on the 9th of February, and crossed the Dymel on the 11th in four columns. The leading column, under the Marquis of Granby (which included General Conway's division),† marched on Volkmissen, where they were cantoned that night.

The Hereditary Prince was ordered to advance

The following is a list of the British Regiments employed in Germany from the same state (exclusive of Guards):—

Major-Gen. Hodgson, 5th.	Major-Gen. Erskine, 25th.
Major-Gen. Barrington, 8th.	Major-Gen. Griffin, 33rd.
Lieut.-Gen. Bockland, 11th.	Lieut.-Gen. Stewart, 37th.
Lieut.-Gen. Napier, 12th.	Lieut.-Gen. Carr, 50th.
Lieut.-Gen. Kingsley, 20th.	Major-Gen. Brudenel, 51st.
Lieut.-Gen. Boscawen, 23rd.	Lieut.-Colonel Keith, 87th.
Lieut.-Gen. Cornwallis, 24th.	Lieut.-Colonel Campbell, 88th.

* West of Lunen.

† The "History of the Grenadier Guards," vol. ii. p. 175, gives the troops under Lord Granby's command at this period, which includes, among others, General Conway's division, of which the 50th Regiment formed part of the 2nd Brigade.

The same work says, Lord Granby "with the Guards and the rest of the British troops taking the lead to Treyza."

A letter from Lord Granby of February 11th, 1761, says he was through the pass of Waldeck towards Fritzlar, and drive in the enemy's posts on the Edler, while General Sporcken's column was to have done the same on the Werra, where they destroyed the magazines at Langensalza, the remaining column being under Prince Ferdinand.

This enterprise was so well planned and so unexpected, that it succeeded at all points, and the Duc de Broglio and his army were driven back to the River Mayne; before their retreat, however, they left large garrisons in Gottingen and Cassel. To the siege of the latter place Prince Ferdinand now applied himself, opening the trenches before it on the 1st of March. He encountered great difficulties from the brave defence and the inclemency of the weather. The French army under De Broglio soon rallied, and, resuming the offensive, they beat the Hereditary Prince near Grunberg on the 21st of

in command of the right column, while the position he assigns to the columns of the Hereditary Prince and General Sporcken would make his column central. The explanation appears to be that Lord Granby's was the right column in crossing the Dymel river; the column of the Hereditary Prince crossing that river in rear of Lord Granby, and having much further to go, it would be a long time before the Hereditary Prince got on his right. This would make the former both the leading and the right column.

The same letter says that Lord Granby's force was cantoned that night at Volkmissen, and was to attack Durenberg next morning (in the direction of Cassel), and on the 19th I find his column approaching Ziegenhayn (where a garrison had been left by the enemy), which is on the road to and near Treysa.

March, and compelled Prince Ferdinand to raise the siege of Cassel after twenty-seven days.

Both armies then resumed the positions they had occupied during the winter, the allies recrossing the Dymel on the 1st April.

The allied army had, however, in the meantime destroyed a great many French magazines, which prevented the latter from renewing hostilities till the end of June, at which period Marshal Soubise's army advanced from the Rhine and Marshal De Broglio's from the Fulda, both armies manœuvering to join their forces and to give battle to Prince Ferdinand, who made no effort to prevent the junction, confident in the strength of his position.

On the 25th of June Lord Granby marched to Soest, and at 12 o'clock that night he was detached with the two battalions of British Grenadiers, two battalions of Highlanders, and Colonel Harvey's brigade of cavalry to Wippringhausen. He was joined next day by Hodgson's (5th), Napier's (12th), Cornwallis' (24th), and Stuart's (37th) Regiments, and on the evening of the 27th he took up a position on the heights of Rhuden to cover the advance of the main army, which marched at 12 o'clock that night from Werle. In addition to the troops already enumerated, Lord Granby had now under his command 2 battalions of Mansberg's Regiment, 10 six-pounders, 5 squadrons of Elliott's Dragoons, and the Hessian Chasseurs.

On the 5th of July Prince Ferdinand endeavoured to bring a portion of the enemy to action.* With this view, the Hereditary Prince was ordered to attack their left, and the battalions of Grenadiers and Highlanders, Elliott's Dragoons, and all the Hussars their right. The enemy, however, proved to be too strongly posted, and the troops were withdrawn.

Shortly afterwards General Sporcken was compelled to withdraw his troops from the Dymel (or Diemel) river,† and Marshal De Broglio took possession of Paddeborn, and was reported to be marching on Lippstadt. On the 6th, however, he effected his junction with Marshal Soubise through the Forest of Testenberg, near Soest.

Ten days were now lost by the united armies in reconnoitring Prince Ferdinand's position, and they do not appear to have very cordially co-operated.

They occupied a strong position to the south of the River Lippe, their left stretching towards Soest, and their right reaching to the heights of Rhuden.

Prince Ferdinand awaited their attack; his left, under Lord Granby (including General Conway's division), in the narrows between the rivers Ahse and Lippe, was posted on a hill between the villages of Kirch-Denkern, and Velinghausen, commanding

* Probably a portion of Marshal Soubise's force, which arrived first.

† De Broglio is said to have beaten Sporcken and taken 800 prisoners and 19 cannon.

the road between Ham and Lippstadt. Near this position the Ahse river is joined by the Salzbach, a small deep river, behind which ran an eminence on which the centre was posted (at first held by the Prince of Anhalt), while the extreme right, behind Werle, was held by the Hereditary Prince, his front protected by almost impassable ground.

About 5 p.m. on the 15th July firing was heard on the left, and the troops were disposed to resist an attack; about 6 o'clock Marshal de Broglio advanced against Lord Granby's position, while Marshal Soubise held a large force in readiness should this attack prove successful.

Lord Granby had orders to maintain his position to the last, and the British troops were posted on the wooded heights of Velinghausen—the enemy drove in Lord Granby's outposts, but were unable to reach these heights, where the fight was continued with "indescribable bravery" by the British troops until 10 o'clock at night, when De Broglio's troops were withdrawn.

The attack was renewed at 3 a.m. next morning (16th), but General Wertgenau's command had now reinforced Lord Granby, taking post on his extreme left, and the Prince of Anhalt's force was also ordered to take up a position on his right, where a bridge over the Ahse was in the possession of the enemy, General Conway's division taking his place behind the Salzbach, between Hohenover and Ilinghen. To the

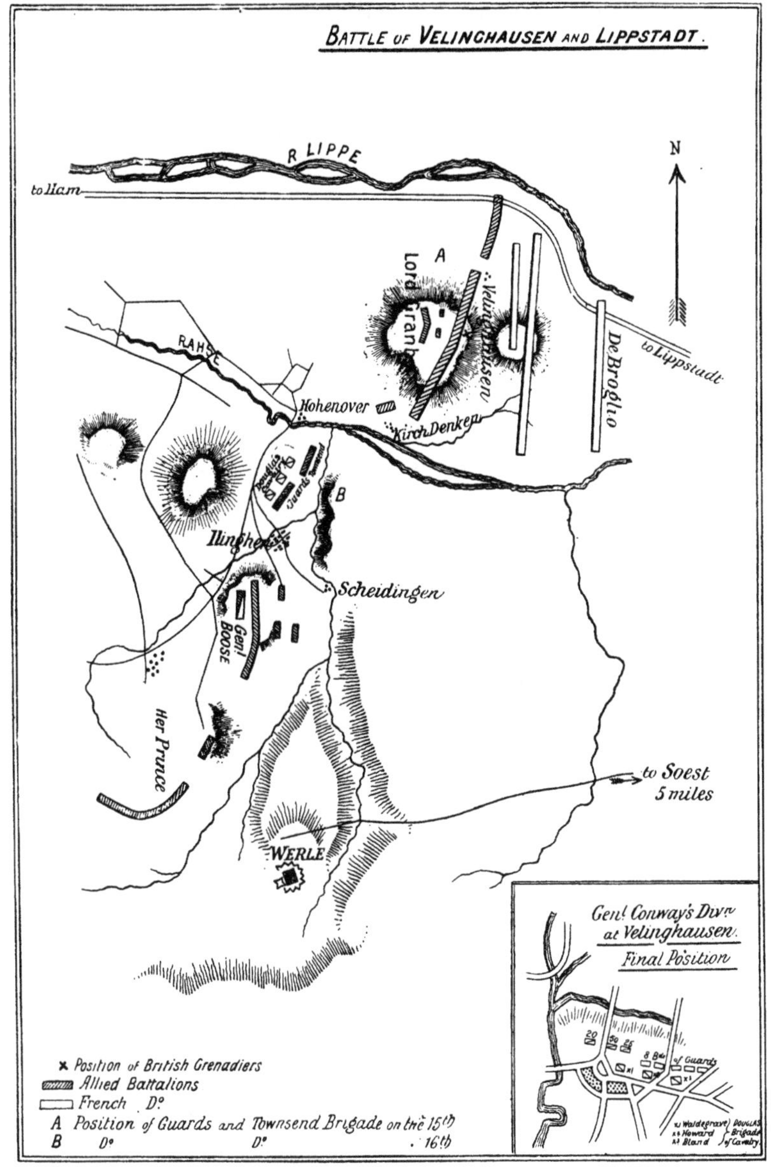
BATTLE OF VELINGHAUSEN AND LIPPSTADT.
R LIPPE
N
to Ham
A
Lord Granby
Velinghausen
De Broglio
to Lippstadt
RAHSE
Hohenover
Kirch Denken
Guards
B
Ilingen
Scheidingen
Genl BOOSE
Her Prince
to Soest
5 miles
WERLE
Genl Conway's Divn
at Velinghausen.
Final Position
20
50
25
8 Bde of Guards
Waldegrave
Howard
Bland
DOUGLAS Brigade of Cavalry
× Position of British Grenadiers
Allied Battalions
French Do
A Position of Guards and Townsend Brigade on the 15th
B Do Do 16th

right of General Conway, General Boose held the heights of Wambeln between his position and that of the Hereditary Prince.

The enemy redoubled their attack on the left of Lord Granby's position, held by General Wertgenau (who was reinforced about 7 o'clock by six battalions from General Sporcken's force on the other side of the Lippe) without gaining any advantage.

About 9 o'clock it was discovered that the enemy were endeavouring to place some cannon on an eminence, opposite to Lord Granby's position, which would have done much damage. An attack was therefore ordered, consisting of the two battalions of British Grenadiers, Keith and Campbell's regiments, and Hanoverian troops; these were ordered to march through a thick coppice and attack the enemy's left, while Napier's regiment, supported by German troops, were to attack in front.

The enemy's right gave way, and the Grenadiers gaining the heights found them moving off. The first battalion fell in with them, killing many and making a great number of prisoners. Maxwell, at the head of the Grenadiers, took the whole regiment of De Rouge. The enemy must have lost 5,000 men in the two attacks.

Colonel Beckwith at the head of the British Grenadiers pushed on about two miles in pursuit of the enemy.

The enemy now attacked the centre, held by General Conway's division, made themselves master of the bridge and village of Scheidingen, and made seven different attacks on a redoubt that commanded the exit from that village, which were always repulsed.

The Irish brigade (in the service of France) made an unsuccessful attempt to force the passage of the rivulet* to the left of Scheidingen, which was held by three battalions.

As the enemy's attack had failed at all points, Marshal Soubise's force took no part in it, and the day ended by a retreat of the French armies.

Prince Ferdinand at once ordered a general advance of his troops, which speedily turned the retreat into a flight. They are reported to have lost 5,000 to 6,000 killed, wounded, and prisoners, with 9 pieces of cannon, and 6 colours.†

The loss of the 50th Regiment consisted of 2 privates wounded and one taken prisoner, but this does not include the portion of the regiment incorporated with the 1st Battalion of British Grenadiers, whose losses would be shown with that corps.

The two armies separated and retired to their former positions on the 26th, and Prince Ferdinand

* This refers to the rivulet running by Ilingen, and the three battalions would be guards from General Cæsar's brigade. See plan of position of Conway's brigade.

† French accounts only admit a loss of 2,400.

followed the army of Marshal De Broglio which marched towards Paddeborn on the 27th.

He succeeded in heading him at Hameln on the Weser, while the Hereditary Prince threatened his rear from the direction of Warbourg.

After repeated skirmishes with varying success, the French position on the Weser was turned by a successful attack of General Luckner on Dassel, on the 14th of August, and on the 18th of that month they retreated across that river at Hoxter, in the direction of Eimbech.

Hoxter was taken possession of by the allies on the 21st, and on the 24th Prince Ferdinand, at the head of Lord Granby's Division, with all the British troops except the Guards, proceeded by forced marches towards the Dymel river, and fired all the enemy's forts in that direction; * then crossed the river and encamped within six leagues of Cassel.

On the 17th September the Hereditary Prince, having driven back Marshal Soubise's army, crossed the Dymel at Warburg and advanced to Wilhelmstadt, where he was joined on the 20th by Prince Ferdinand's army.

In the beginning of November an attempt to cut off a French column under Marshal Chabot, at Escherhausen, in which General Conway's division

* Particularly Drengenbourg on a branch of the Lippe river where 300 prisoners were taken.

held an advanced position, was frustrated by the overturning of General Hardenberg's pontoon train.

The following extract from the *London Gazette* will give some idea of the harassing nature of the duties that fell to the British troops at this period.

"British Camp at Worwolde, November 8th.

"Since the 3rd of this month we have been continually under arms. Our first operation was the dislodging of the French corps encamped at Escherhausen, when we were assigned to take them in rear, while other corps were to fall on their front and flank. We drove in their advanced posts, but they would not await an attack. We were then ordered to march on towards Eimbech. When we got to the neighbourhood of that place we found the Hereditary Prince engaged in a cannonade with the enemy's army; but their strong position made it impossible to make any impression. We remained there that night, and next day, after which we were ordered to march to this place, Worwolde, which we did in the night, through a heavy fall of snow and over almost impassable roads. Our tents were but just pitched and the camp formed, when we heard a report from outposts that the enemy were coming down upon us in a strong body. Our troops thereupon instantly formed, advanced, and attacked the French with the greatest spirit and drove them back almost to their very camp, and we had the good fortune to have our General-in-Chief, Prince Ferdinand, a spectator of what occurred.

"The next morning Lord Granby was again attacked before he could march to follow the main army, which was endeavouring to work round to the rear of De Broglio, but he again repulsed the enemy with considerable loss. De Broglio, now finding the Prince to be on his flank and rear, saw he must either risk an action or retire, and deciding on the latter, he quitted Eimbech and the adjacent country on the 9th."

Lord Granby reporting on this action at Worwolde says:—

"I can never enough admire the extraordinary spirit and gallant behaviour of the troops under my command after the extraordinary fatigue they had undergone.

"The corps engaged were the British Grenadiers and the Highlanders under Lord F. Cavendish, Lieutenant-General Schules, and Major-General Pincier's brigades of infantry, and Colonel Harvey's brigade. Schules, Cavendish, and Johnston's brigades under the command of Lieutenant-General Conway,* and joined with him are the 5th at Wentzer.

"I must particularize Maxwell's chasseurs."

The troops moved into cantonments for the

* At this period Lieutenant-General Conway appears to have ceased to command Townsend's Brigade, and from this date we find an entirely new disposition of the brigade.

winter on the 12th of November; General Conway at Eimbech, with Hodgson's, Bockland's, Barrington's, and some German regiments.

The 50th and most of the British battalions were cantoned in the bishopric of Osnaburgh.

Early in the year 1762 the corps of General Luckner and the Hereditary Prince were engaged with the enemy, but the British troops did not assemble till June 4th, when they joined the *corps d'armée* of General Sporcken near Blomberg.

On the 18th of that month the whole allied army was united at Brackel.

The 50th Regiment was brigaded with the same troops as in the previous year, but the division was now commanded by General Sporcken, and the brigade which occupied the right of the second line of the division was commanded by General Waldgrave.

On the 20th of June Lord Granby's corps, which comprised both battalions of the British Grenadiers, under Colonel Beckwith, advanced to Warburg on the Dymel river.

On the 21st they crossed the Dymel and reconnoitred the enemy's left towards Volckernissen, returning the same evening, and about the same date the castle of Zappaburg on the enemy's right front and all the passes of the Dymel river were seized.

On the 22nd the whole French army, under

Marshals D'Estrees and Soubise, took up a strong position to cover Cassel; their right near Grevenstein and their left protected by the wooded heights of Meizenbrecksen.

Here Prince Ferdinand determined to attack them. With this view General Luckner, who had been posted on the Leine river watching Gottingen, broke up his camp on the night of the 22nd; passed the Weser at Bodenfelde at six in the evening on the 23rd; encamped for the night at Gotsbuhren, and marching again at 3 a.m. by the woods and castle of Zappaburg, he took up a position at 7 a.m. on the 24th on the enemy's right flank, between Mariendorf and Underhausen.

The main body of the army passed the Dymel in seven columns, to their left of Liebenau, at 4 a.m. on the 24th June. Two of these columns, under General Sporcken, consisting of Hanoverian troops, were ordered to form up on and attack the enemy's right flank, while General Luckner, who was already posted there, attacked it simultaneously from the rear.

Prince Ferdinand himself commanded the remaining five columns, which comprised among others twelve battalions of British infantry, one of which must have been the 50th Regiment.* This force,

* The two columns commanded by General Sporcken consisted of 12 Hanoverian battalions and cavalry. (Operations of Allied Army.)

I have not been able to obtain details of the composition of

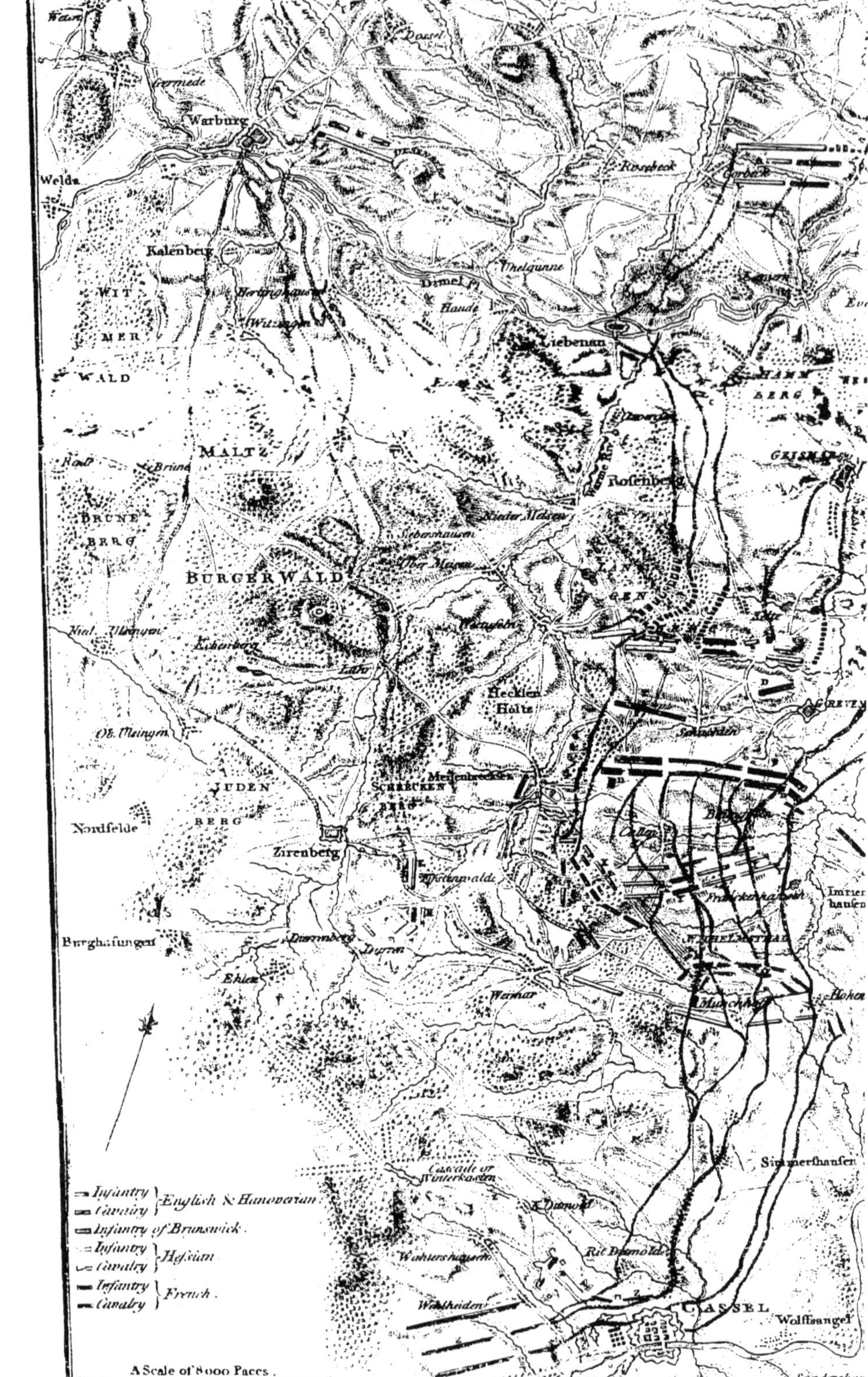

Warburg
Welda
Kalenberg
WIT
MER
WALD
MALTZ
BRUNE
BERG
BURGERWALD
JUDEN
BERG
Nordfelde
Zirenberg
Burghafungen
Liebenau
Rosenberg
Heckle
Holtz
SCHRECKEN
BERG
CASSEL
Wolffsangel
Simmerfhausen
Infantry
Cavalry
English & Hanoverian
Infantry of Brunswick
Infantry
Cavalry
Hessian
Infantry
Cavalry
French
A Scale of 8000 Paces

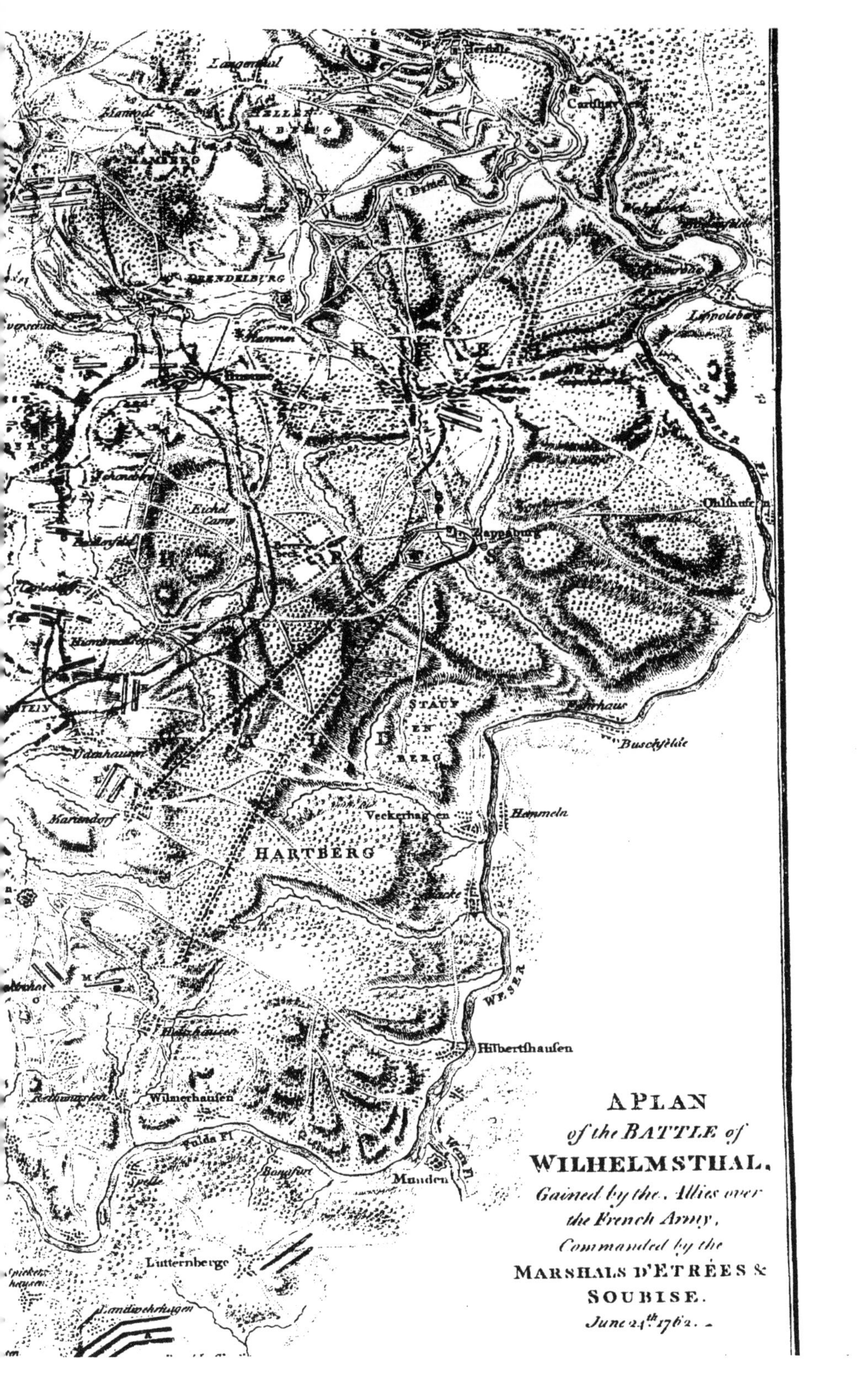
Langenthal
Carlshaven
HELLEN BERG
Lippoisberg
WESER Fl.
Ohlshausen
Zappaburg
Eichel Camp
Hartberg
STAUFEN BERG
Buschfelde
Veckerhagen
Hemmeln
HARTBERG
WESER
Hilbertshausen
Wilmerhausen
Fulda Fl.
Munden
Lutternberge
Landwehrhagen
A PLAN
of the BATTLE of
WILHELMSTHAL,
Gained by the Allies over
the French Army,
Commanded by the
MARSHALS D'ETRÉES &
SOUBISE.
June 24th 1762.

which was intended to attack the enemy in front, formed up to their right of the little village of Kelz, to the north-west of Grevenstein, while sixteen squadrons of cavalry took part at Geismar, in rear of their left.

Lord Granby's force, which included the British Grenadiers, passed the Dymel, in three columns, between 2 and 3 a.m., and were ordered to attack the enemy's left flank from the direction of Zirenberg.

These operations were carried out with such success that the columns had taken up their respective positions before the enemy had any apprehension of being assaulted.

The attack of Generals Luckner and Sporcken on the enemy's right, assisted by the cavalry, at Geismar, was also completely successful, and that flank was driven back in confusion.

Prince Ferdinand now attacked in front, while the Marquis of Granby moved to the rear of the enemy's left flank at Zirenberg, and advanced through Fursstenwalde, threatening to cut off the retreat of the whole army.

The two marshals, seeing themselves attacked in front, flank, and rear, abandoned all their equipages

Prince Ferdinand's force of five columns; but as it comprised twelve British regiments it must have included the 50th Regiment, as, deducting Lord Granby's command, it would take all the British infantry then in Germany to provide twelve battalions.

and used the utmost expedition in retiring to the heights of Wilhelmsthal.

The French General, M. de Steinville, perceiving that Lord Granby's force cut off the retreat of part of the army, threw himself into the woods between Meizenbretsen and Wilhelmsthal to cover their retreat. Here he was at once attacked by Lord Granby, and after a gallant resistance the whole of his force was either killed, taken, or dispersed, except two battalions which managed to escape.

The enemy now retired to their entrenched camp under Cassel, being closely pursued by the allies until within reach of the cannon of the town, and Prince Ferdinand encamped that night between the heights of Wilhelmsthal and the town.

The loss of the enemy in killed and wounded was not known, but their own estimate is said to be 900 men, in addition to which 1 standard, 6 colours, 2 cannon, and 2,732 prisoners were taken.

The following is the report on Lord Granby's command:—

"Lord Granby acquitted himself upon this occasion with his usual intrepidity and good conduct, and greatly contributed to the victory. All the troops in general behaved with uncommon spirit, but particularly the 1st Battalion of Grenadiers, belonging to Colonel Beckwith's brigade, who distinguished themselves greatly.

"The loss of the 1st Battalion of British

Grenadiers, now under the command of Colonel Welsh, was 3 privates killed, 40 wounded, and 3 prisoners.

"The total loss of the allies was 10 officers, 33 non-commissioned officers, and 753 privates killed, wounded, and missing."

The remainder of this year was occupied in desultory operations, principally in the neighbourhood of the Rivers Ohm and Lahn, in which the allied army had generally the advantage.

Early in July, Prince Ferdinand determined to drive the enemy from a strong position which they had taken up at Homburg, and a force under Lord Cavendish and Lord Granby was selected for this service.

About the 1st of that month Elliott's cavalry drove in the enemy's outposts, when they at once struck their tents and made preparations to retreat, covered by their cavalry. Lord Granby perceiving this, ordered his cavalry to advance, which they did, falling on the enemy's rear with the greatest ardour and success; but the cavalry of the retreating side turning about, attacked Elliott's dragoons while scattered in pursuit, and they were only saved by the promptness with which "The Blues" came to their rescue.

The situation of the two regiments remained very critical, but they kept the enemy at bay till the infantry came up, when the French retreated in haste, "the British Grenadiers and Highlanders following them with their usual ardour."

On the 15th July, Prince Ferdinand marched from Wilhelmsthal to Hoff, and all the enemy's posts on the Fulda as far as Melsungen were attacked. In particular, the corps under the Marquis of Granby advanced to the Edler, leaving their tents standing. The British Grenadiers, the Highlanders, and Bland's Regiment forded that river between Nieder Melric and Felzberg, while General Freytag passed over the bridge of Felzberg; but the enemy having been considerably reinforced, the troops had to retire.

During that month Prince Ferdinand arranged an attack on Prince Xavier's corps at Luttenburg (east of Cassel and south of the Fulda).

On the night of the 23rd three columns crossed the Fulda. The Grenadiers on the left began the attack about 4 a.m., crossing the river under a severe fire of cannon and musketry, followed by the rest of the infantry, though the water was above their waists. The affair was sharply disputed for some time, but at length the enemy were driven out of four palisadoed redoubts, and from all their entrenchments, and retreated to Cassel, leaving one regiment of cavalry and two of Grenadiers as prisoners.

The French general De Stainville, having marched out from his entrenched camp in the Krazenberg to cover their retreat, Prince Ferdinand marched on it, took possession of it, and destroyed all the works and redoubts there.

The enemy's loss was very considerable, and the prisoners amounted to 1,100, besides 13 pieces of cannon and 2 standards.

The loss of the allies did not exceed 200 men.

On the 18th and 19th of September an obstinate conflict took place for the possession of the castle of Amonebourg.

A desperate struggle took place across the River Ohm. The troops on both sides being constantly relieved as soon as their ammunition was exhausted.*

The attack continued for 14 hours without a moment's interruption, and though 50 pieces of cannon were employed, their execution was confined to about 400 paces.

The loss of the allies was computed at about 800.

The French acknowledged a loss of 300 killed and 800 wounded.

The castle surrendered to the French on the 22nd, after a practical breach had been made.

On the 16th September Prince Ferdinand opened the trenches before Cassel, which had been blockaded

* Lord Granby writing about this affair under date of October 2nd says:

"History furnishes no account of so obstinate a dispute. The redoubt was exposed to the enemy's cannon at 300 paces and small arms at 30, there were 500 dead in the redoubt, and the troops that came late used the dead to form a parapet. Maxwell's battalion of British Grenadiers were engaged."

for some time by the Prince of Brunswick, and which surrendered on the 1st November, after two sallies (on the 16th and 22nd).

On the 14th of November a messenger arrived with the information, that preliminaries of peace had been signed on the 3rd instant at Fontainebleau. A suspension of hostilities took place on the 15th, and the two armies began to file off towards their respective quarters; the British troops moving towards the bishopric of Munster.

The thanks of the House of Commons were communicated to the troops on the 13th of January, 1763. Lord Granby, who had returned to England, also sent a letter of thanks to the British troops.

On the 15th of January, Carr's and Brudenel's regiments were ordered up, in consequence of a dispute about pay—on account of which the Legion Britannique were in a state of mutiny—which was promptly quelled.

The first division of British troops began their march through Holland to Williamstadt on the 25th of that month; where the embarkation of troops in two divisions began on the 21st of February.

The 50th Regiment remained in cantonments on the frontiers of Holland till March 1763, when they returned home, and were quartered at various stations in Ireland.

CHAPTER III.

CORSICA.

THE 50th Regiment remained in Ireland till the latter part of 1772, when it embarked for Jamaica, where it arrived at the beginning of 1773. A light company was added to its establishment on the 21st September, 1771. It remained at Jamaica till the beginning of 1776, and was then ordered to North America, where it was broken up, and the whole of of the men fit for duty were drafted to reinforce other regiments. The staff was sent to England to recruit to their full strength, and arrived at Salisbury for this purpose in November 1776.

On the 3rd July, 1778, the regiment, now completed, embarked in H.M. SS. "Centaur," "Vengeance," "Defiance," "Thunderer," and "Vigilance," and were employed as marines in an indecisive attack made by Admiral Keppel on the French fleet under Count D'O. Villiers off Ushant. On the return of the fleet they were disembarked and marched to Exeter.

E 2

The corps was directed to assume the title of the 50th or West Kent Regiment on the 31st of August, 1782.*

On the 24th June, 1783, the regiment was reduced from ten to eight companies. It embarked for Gibraltar on the 3rd August, 1784, under the command of Lieutenant-Colonel Edmeston, and landed on the 21st of the same month. It remained at Gibraltar until January 1794,† when it embarked and joined the fleet under Admiral Lord Hood in Hyeres Bay.

Lord Hood having heard, after he had evacuated Toulon, that the Corsicans were dissatisfied with the French Government and straitened for food, and having 4,000 or 5,000 soldiers on board under

* By the "Additional Force Act" of 29th June, 1804, the county allotted to the 50th Regiment was Gloucestershire.

† The Regimental records say the 50th remained at Gibraltar till January, 1793, when they sailed for Corsica, but the following extract from a despatch of Admiral Lord Hood, dated March 18th, 1794, makes it evident that they sailed from Gibraltar to Hyeres Bay.

Admiral Hood writes: "When the 50th and 51st Regiments joined H.M. Fleet in Hyeres Bay from Gibraltar," &c., &c.

And in a letter from Lieutenant-General Dundas, dated St. Fiorenzo, February 21st, 1794, he states that the 50th (with other regiments) sailed from Hyeres on the 5th February, 1794; and as the fleet did not go into Hyeres Bay till December 19th, 1793, the date of leaving Gibraltar must have been January, 1794. They could not have taken from January, 1793, to February, 1794, on the voyage.

General Dundas, determined to make an attempt to expel the French from the island.

With this view the fleet sailed from the Bay of Hyeres on the 24th January, 1794.

On the 5th February, 1794, the troops, consisting of the 2nd Battalion of the Royals, 11th, 25th, 30th, 50th, and 51st regiments—in all about 1,400—sailed in transports escorted by 2 ships of the line and 2 frigates. They anchored in the Bay of Martello, near St. Fiorenzo, on the 7th of that month, landed the same evening, and took possession of a height that overlooked the Tower of Martello.

On the 8th the fleet opened fire against the tower, and the artillery battered it from the height; eventually a red-hot shot set fire to some bass junk with which a parapet was lined, and the garrison surrendered.

It was next decided to attack the Convention Redoubt, which was of a long narrow form, occupying the summit of a detached height about 250 feet above the sea, mounting 21 pieces of heavy ordnance, and said to be the key to the bay.

On the 16th February two batteries, each of three pieces, were, with incredible labour, dragged by the sailors up a steep and dangerous mountain path, to a detached hill 700 feet above the sea, commanding and enfilading the redoubt, and opened fire upon it.

The following evening an assault was decided on. The troops marched in three columns; and having nearly equal distances to go, started at the same hour

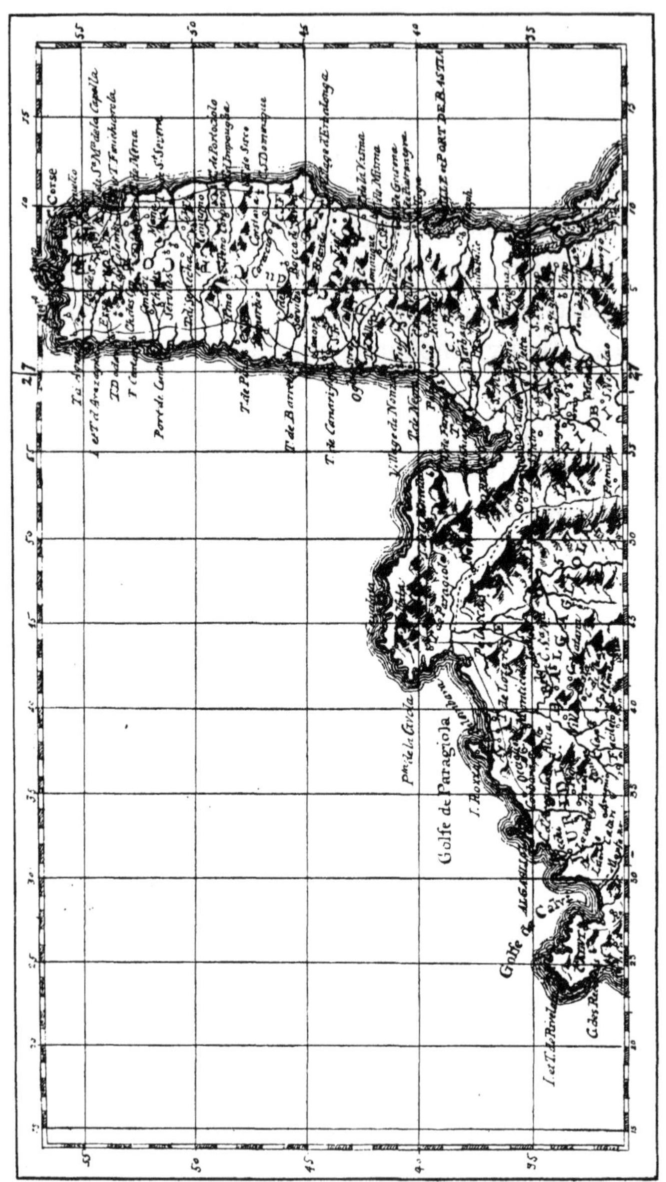

(8·30 p.m.) in order to arrive at the enemy's works a little after the rising of the moon. Lieutenant-Colonel Moore on the right, with the 2nd battalion of the Royals and the 51st Regiment moved on the advanced fronts of the redoubt.

Lieutenant-Colonel Wauchope with the 50th Regiment advanced towards its centre.

Captain Stewart with the 25th Regiment, keeping close along the shore, was directed to enter on the left.

Notwithstanding that the whole of the ground over which the troops marched was rocky, rough, and covered with thick myrtle, they approached the redoubt, under a heavy fire, without the enemy being aware of their progress. They arrived at their respective posts nearly at the same time, rushed into the works, prevented more than two or three discharges of cannon, and with their bayonets drove the enemy down the steep hill which formed the rear of the works.

General Dundas, in his report on this attack, says: "The conduct of Lieut.-Colonel Moore, of the several commanding officers, and of all the officers and soldiers, was firm and judicious, and merits every commendation."

The loss of the 50th Regiment was—1 private killed and 5 wounded.

The total loss of the enemy was reported to have been—10 officers and upwards of 100 privates killed and wounded, and 60 prisoners, out of a garrison of 550.

The enemy, being afraid of their retreat being cut off, abandoned a strong post on the heights of Forneli, 400 yards off, and retired to Fiorenzo, hauling up their 2 frigates, which were afterwards sunk and burnt.

On the 19th February Fiorenzo was evacuated, and immediately occupied by our troops. The enemy retreated towards Bastia, abandoning several strong posts on the way, and taking up their position half-way to Bastia, which is about 11 miles from Fiorenzo, and separated from it by a ridge of mountains. The principal defences of Bastia were four detached masonry forts, which covered the land side, and several batteries facing the sea. Our advanced posts being now within 2 miles of Bastia, it was reconnoitred by General Dundas, who considered it impracticable with the force under his command. Admiral Lord Hood, however, determined to attack it from the sea; and a certain proportion of the troops, guns, ammunition, and stores having been handed over to him, the remainder of our troops were withdrawn from the mountains, the climate of which was very cold, towards Fiorenzo. The 51st and 69th Regiments were encamped 3 miles from the town, the 50th half-way between them, and the rest of the garrison, consisting of the 1st and 18th, in the town.

On the 15th April, reinforcements having been received, these troops, which were now under the command of General Daubert, took post on the heights above Bastia, leaving a small garrison at Fiorenzo, but they appear to have taken no active part in the siege, which surrendered to Lord Hood on April 22nd.

On the 31st May Lieutenant-General Stewart, who had recently arrived from England, and taken over the command of the troops, sailed in H.M.S. "Illustrious," to reconnoitre Calvi.

On the 13th June 1,450 soldiers, exclusive of officers, embarked at Bastia about 8 a.m. and sailed for Martello Bay, St. Fiorenzo, at 4 p.m., convoyed by the "Agamemnon," "Dolphin," and "Gorgon," under the command of Captain Horatio Nelson; they arrived on the 15th, and took General Stewart on board.

At 5 p.m. on the 16th the "Agamemnon," "Dolphin," "Lutine," and 16 transports, with the troops under General Sir C. Stewart (which included the 50th Regiment), sailed for Calvi. They anchored on the 17th opposite a little inlet called Porto Agro, about 4 miles west of Cape Ravelata, and 3½ miles from the town of Calvi. The inlet was full of sunk rocks, 20 feet from the shore, with deep water between and commanded by steep cliffs.

At 7 a.m. on the 19th all the troops landed, having 6 field pieces, which the sailors with great labour

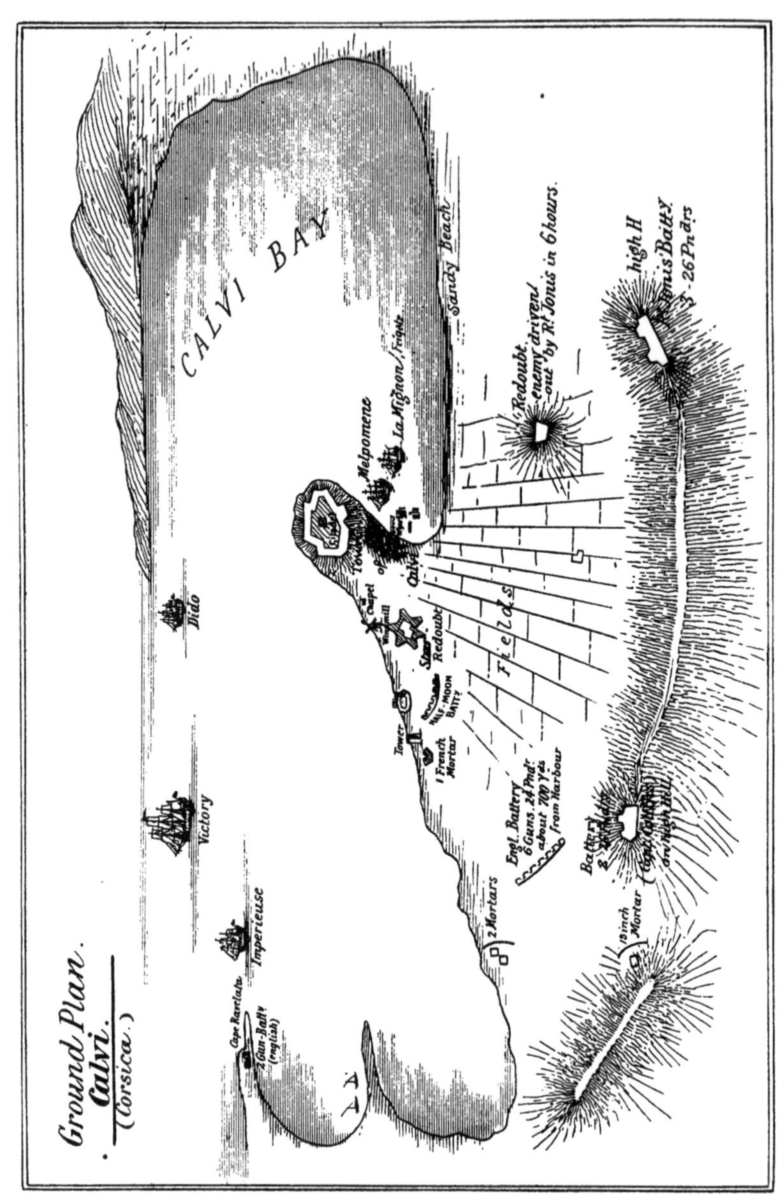
Ground Plan.
Calvi.
(Corsica.)
CALVI BAY
Sandy Beach
Redoubt.
enemy driven
out by Rt Jonis in 6 hours.
high H
Rt Jonis' Batty.
3 26 Pndrs
Melpomene
La Mignon, Frigate
Dido
Victory
Imperieuse
Cape Revellata
2 Gun Batty
(english)
Fields
Redoubt
Chapel
Windmill
Tower
Half-Moon Batty
1 French Mortar
Engl. Battery
6 Guns. 24 Pndr.
about 700 yds
from Harbour
2 Mortars
Battery
Mortar

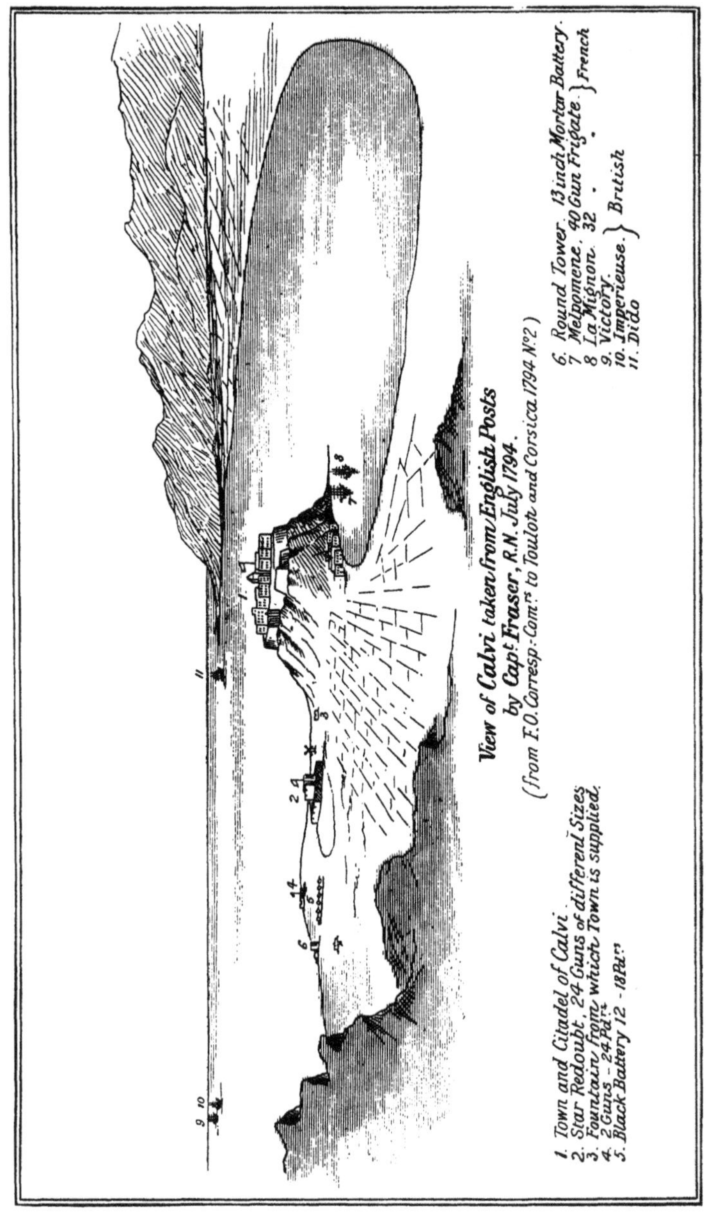
View of Calvi taken from English Posts
by Capt. Fraser, R.N. July 1794.
(from F.O. Corresp: Comrs to Toulon and Corsica 1794 No. 2)
1. Town and Citadel of Calvi.
2. Star Redoubt. 24 Guns of different Sizes
3. Fountain from which Town is supplied.
4. 2 Guns - 24 Pdrs
5. Black Battery 12 - 18 Pdrs
6. Round Tower. 13 inch Mortar Battery.
7 Melpomene. 40 Gun Frigate. } French
8 La Mignon. 32 " "
9. Victory.
10. Imperieuse. } British
11. Dido

dragged up the steep hills. In the afternoon 257 sailors landed from the fleet. All hands were now busily employed getting guns, stores, &c., up the hills and making batteries.

Reinforcements consisting of 180 of the Royal Louis Regiment (a French emigrant corps) landed on the 22nd, and the 18th Regiment and the flank company of the 69th on the 26th.

At 1 p.m. on the 27th the enemy made a sortie and attempted to turn both flanks; they advanced under cover of a heavy cannonade, and were at first partly successful against our Corsican allies, killing their colonel and driving them back, but they soon rallied and regained their position, and the French were eventually driven back into Calvi.

The enemy's post consisted of the following batteries: Monachesco, about 2,200 yards south-west of the town, mounting one 18-pounder, four or five 6 and 8 pounders; Mozello, 900 yards from the town (west face with cavalier in centre), two heavy guns towards the land and two towards the sea; Fountain battery, on the shoulder of the hill between Mozello and San Francisco, consisting of six 18-pounders landed from the "Melpomene," and a fascine work; San Francisco on the rock washed by the sea, three 18 and 24 pounders (brass guns).

On the 7th July a battery was completed within 750 yards of the Mozello. At 10 a.m. it opened fire, and at 3 p.m. the enemy abandoned the Monachesco, which the Corsicans took possession of, but on the 8th a very heavy fire destroyed two of our 24-pounders, damaged two 6-pounders, and greatly shook the works; while a shell burst in the centre of the battery in which the general and 400 men were and blew up the battery and magazine. These were repaired at night and the damaged guns were replaced by 18-pounders.

On the 9th the guns of the Fountain and Mozello batteries were silenced.

On the 10th there was a small breach in the Mozello. On the 12th and 13th a heavy fire from San Francisco destroyed several of our guns, but the Mozello was now much breached. On the 15th scaling ladders were provided, and on the 18th, the breach in the Mozello being now very large, everything was ready for attack, and the following disposition was made:

Colonel Wemyss, with the 18th Regiment and 2 field pieces (manned by 30 seamen) was to proceed by the left of our six-gun battery, and with fixed bayonets to take possession of the Fountain battery, then to direct fire of his two pieces on San Francisco. On this signal our three-gun battery and 2 field pieces which were with Colonel Moore, were to fire on the Mozello and clear the breach. Carpenters were to go before and cut down the palisades; some light companies under Major Brereton were to keep to the right of the Mozello, to cut off the retreat of the garrison. Colonel Moore was to command the storming party of 400 men, supported by the 51st Regiment, the 50th Regiment to the right.

From the official report of Lieutenant-General Sir C. Stewart, the above disposition does not, however, seem to have been adhered to. It appears from his reports that on the 18th of July a breach appearing to be practicable on the west side of the Mozello, a disposition was made for a general attack upon the outworks under cover of two batteries to be erected during the night. From the zeal of Lieutenant-Colonel Wauchope and the great exertions of the 50th Regiment under his command, the battery which he undertook to construct within 300 yards of the Mozello was completed by an hour before daybreak, without discovery, and armed at daylight; a signal gun was then fired from it for the troops to advance. Lieutenant Newhouse of the Royal Artillery with 2 field pieces covered the approach, and the 2nd Battalion of the Royals, under the command of Lieutenant-Colonel Moore, with the 51st and 30th Regiments, proceeded with a cool steady confidence and unloaded arms towards the enemy, forced their way through under a smart fire of musketry, and, regardless of live shells flung into the breach, or the additional defence of pikes, stormed the Mozello; while Lieutenant-Colonel Wemyss, with the Royal Irish Regiment and 2 pieces of cannon under Lieutenant Lemoine, Royal Artillery, equally regardless of opposition, carried the enemy's battery on the left, and forced their trenches without firing a shot. The General now offered terms to the garrison, which were declined; but the possession of these important works of the enemy, which were held under the heaviest fire of shot, shell, and grape, enabled fresh batteries of 13 guns, 4 mortars, and 3 howitzers to be thrown up within 600 yards of the town. These opened fire on the 27th, with such good effect that the enemy were unable to remain at their guns, and 18 hours afterwards proposals for a capitulation were accepted, but the terms were not definitely agreed to until the 10th August.

The general commanding says in his despatch on the above subject, "The spirit, zeal, and willingness with which this army has undergone the greatest labour and fatigue in the most oppressive weather, is hardly to be described."

The loss of the 50th Regiment previous to the attack on Fort Mozello was 1 private killed and 1 wounded, and at the attack on Fort Mozello 1 private wounded. (London Gazette, 2nd September, 1794.)

At the conclusion of the siege, during which the soldiers had suffered greatly from sickness,* the

* A letter from General Sir C. Stewart, dated August 9th, 1794, speaks of the wretched state of the army, which had suffered greatly from sickness and was unable to move by land; and adds that it was absolutely necessary that transport for 1,600 men should be detained to convey them to Bastia. Nelson writes,

British troops, including the 50th Regiment, were removed to Bastia.

Shortly afterwards the crown of Corsica was vested in England, and the 50th Regiment proceeded to garrison Ajaccio, of which Lieutenant-Colonel Wauchope was appointed Governor, and on the 1st May, 1795, a letter from Sir Gilbert Elliott authorised him to call out the militia of the neighbourhood.

In July, 1796, the French, having seized Leghorn, were threatening Elba, and a force, under Commander Nelson, consisting of H.M. SS. "Captain," "Northampton," "Inconstant," "Petterell," "Flora," and 3 smaller vessels, the troops being under Major Duncan, R.A., were sent to take possession of it, and succeeded in doing so without loss.

In October, 1796, the combined French and Spanish fleets temporarily held the command of the Mediterranean, and a powerful force was directed against Corsica by Napoleon Buonaparte. Fortunately they did not succeed in landing together; but the inhabitants having sided with the party that landed, the island was no longer tenable by the small force of British alone, and on the 19th of that month it was evacuated, and the whole of the British troops, including the 50th Regiment, embarked and arrived at Porto Ferrajo, Elba, on the following evening.

under date August 5th, 1794: "We only keep half our seamen at the batteries, yet we have 70 sick, and I sent 30 to the 'Agamemnon.' The troops are worse than ourselves by far.

A letter from Sir G. Elliott, October 26th 1796, reports the complete and successful evacuation of Corsica, and adds, "the enemy had actually arrived on shore, and with a superior fleet at sea, not without reason, I congratulate your Grace, on the entire success of this operation without loss or accident."

In 1797 the 50th Regiment returned to Gibraltar, where it was recruited by drafts from other regiments, after which it sailed for Portugal, arriving there in June of that year. It was quartered in Fort St. Julien, where it continued for two years. It embarked for Minorca in 1799, and eventually joined Lord Keith's portion of the expeditionary force to Egypt from that place in 1800.

CHAPTER IV.

EGYPT, 1801.

An expedition against the French in Egypt having been decided on, the command of it was given to Sir Ralph Abercrombie, K.C.B.

The first part of the fleet appointed to convey this expedition sailed for Minorca on the 3rd November, and the remainder, with Sir R. Abercrombie on board, left on the same date for Malta, where it arrived on the 30th November, 1800. Lord Keith, with the division from Minorca (which now included the 50th), joined the rest of the fleet at Malta on the 14th of December.

The first division of the fleet sailed for Marmorice Bay on the 20th December, and arrived on the 28th. The second division followed on the 21st December, and arrived on the 1st January, 1801.

The sick were landed and encamped, and the troops practised in disembarking, while the expedition waited in this fine harbour for the expected co-operation of troops from India and gunboats from Turkey.

"On the 8th of February a violent thunderstorm arose, which continued for two days. The hailstones were as large as walnuts. The camps were deluged by a torrent two feet deep, which, pouring from the mountains, swept everything before it. The ships in the harbour were in disorder from driving, loss of spars, &c., and the 'Swiftsure' was struck by lightning. Language fails to convey an adequate idea of this tempest." (Wilson's "Egypt.")

The army embarked on the 20th of February, but, owing to the severity of the weather, it was unable to sail until the 23rd of that month.

The expeditionary force was divided into the following brigades:—

Brigade of Guards. Major-General Ludlow.	Coldstream Guards. 3rd Regiment do.
1st Brigade. Major-General Coote.	54th Regiment, 1st Battalion. 54th Regiment, 2nd Battalion. 92nd Regiment.
2nd Brigade. Major-General Craddock.	8th Regiment. 13th Regiment. 18th Regiment. 90th Regiment.
3rd Brigade. Major-General the Earl of Cavan.	27th Regiment. 50th Regiment. 79th Regiment.
4th Brigade. Brigadier-General Doyle.	2nd Queen's. 30th Regiment. 44th Regiment. 89th Regiment.

5th Brigade. Brigadier-General Stuart.	Stuart's Regiment. De Boll's Regiment. Dillon's Regiment.
Reserve. Major-General Moore	23rd Regiment. 28th Regiment. 42nd Regiment. 58th Regiment.
and	
Brigadier-General Baker.	Corsican Rangers.* Flank Companies, 40th Regt. Staff Corps.

Total of rank and file, 15,370.

In addition to the above there were—

Brigadier-General Finch.	12th Dragoons. 26th Dragoons.

The strength of the 50th Regiment, from the War Office returns, was—

2	Majors.		Rank and File.
7	Captains.	458	Fit for duty.
17	Lieutenants.	37	Sick present.
3	Ensigns.	24	Sick absent.
4	Staff.		
28	Sergeants.	591	Total.
11	Drummers.		

Lieutenant-Colonel Rowe commanded the regiment in the absence of Lieutenant-Colonel Walker.

* The Corsican Rangers were raised and commanded by Major Hudson Lowe, late of the 50th Regiment and the son of an officer of that regiment. He is best known as the Governor of Elba during the time when Napoleon was a prisoner. (See Memoirs in Appendix.)

The fleet anchored in Aboukir Bay on the 2nd of March. Unfavourable weather prevented the landing of troops for some days, but on the 7th the shore was reconnoitred, and on the 8th the landing took place.

At 2 a.m. the first division of troops embarked, consisting of the Reserve Brigade, the Guards Brigade, and part of the 1st Brigade, amounting to about 5,500 men. The whole under the command of Major-General Coote.

The 2nd Brigade and the remainder of the 1st were placed in ships near the shore, to be in readiness as a support.

At 9 a.m. the boats anchored about gunshot from the shore; and shortly afterwards the signal for landing was made, and the boats sprang forward, protected on each flank by gunboats.

The French, to the number of about 2,000, were posted on the sand hills in front, in their centre a nearly perpendicular height, on their left Aboukir Castle. As the boats approached the shore, the enemy opened fire on them from the heights, and from Aboukir Castle. The quantity of shot, shell, grape and musketry so ploughed the surface of the water, that it seemed as if nothing on it could live. Several of the boats were struck, some were sunk, and there was a slight confusion; but the majority of the boats, pressing on through the shower of projectiles, forced their way to the shore, when the Reserve leaped out on the beach, forming as they advanced; the 23rd and 40th Regiments rushed up the heights without firing a shot, charging and breaking the two battalions that crowned it. Then, pursuing them to the hills in rear, they captured the position and three pieces of cannon. The 42nd carried the height in front of them, though exposed to the fire of 2 guns and a battalion of infantry, and charged by 200 French dragoons, whom they repulsed.

The Guards Brigade had barely reached the shore before they were charged by the same cavalry, which was defeated, with the assistance of the 58th Regiment. The 54th and Royals in slower boats landed a little later, at the instant that a column of 600 of the enemy were advancing against the left of the Guards. These at once retreated after firing a volley. The French now retreated from the heights, keeping up a desultory fire from the sand hills in rear. The day was over, but at what a cost! 500 brave men had paid the penalty with their lives. The French are reported to have lost 300 men and 8 guns.

General Coote's party now took up a position three miles in advance (towards Alexandria), and the remainder of the troops landed (including the 50th Regiment in Lord Cavan's brigade). In the general orders of the 19th February, a battalion of marines was ordered to land with this brigade, and to take post between the 50th and 79th, but it did not continue to do duty with them.

On the 12th of March the whole army moved

forward, and came within sight of the enemy, who was formed on an advantageous ridge, with his right towards the Alexandria Canal, and his left towards the sea, barring the approach to Alexandria.

It was determined to attack the enemy's position on the morning of the 13th, with a view to turning his right, and on that date a movement was made to the left in two lines, led respectively by the 90th and 92nd Regiments.

The 3rd Brigade, under Lord Cavan (who commanded this column), occupied an advanced position in the left column. After a short advance, and before they were clear of a tope of date trees, the enemy left the heights on which they had been formed, and moved down by their right, pouring a heavy fire of musketry, and opening a destructive cannonade from all their guns on the 92nd Regiment, which led the left column and still continued the advance.

The 90th Regiment, which led the right column, at the same time splendidly repulsed a vigorous charge.

Sir Ralph Abercrombie's despatch says:—

"The troops at once changed their position with a quickness and precision that did them the greatest honour, and were soon in a position not only to face but to repel the enemy."

The army continued to advance in two lines, pushing the enemy with the greatest vigour, until they were compelled to put themselves under the protection of the fortified heights, which form the principal defence of Alexandria.

Sir R. Abercrombie wishing to follow up his success, by carrying the important position that the French had retired to, advanced across the plain, ordered General Hutchinson, with the second line, to move forward to the left, and secure a projecting rising ground, and General Moore to attack the right flank at the same time. The first line remained in the plain rather to the right. While Sir R. Abercrombie was reconnoitring the French from this new position, the British troops were exposed to a most destructive fire.* At length he decided that the position when taken could not be maintained, and at sunset the army was withdrawn. The British loss was about 1,100 killed and wounded.

The 50th Regiment lost: -

Lieutenant Stewart and 5 privates killed.

Ensign Rowe, 1 serjeant, 1 drummer, and 37 privates wounded.

The position now held by the British was very strong, their right was projected for a quarter of a mile on very high ground, and extended to the

* Colonel Wilson in his "Expedition to Egypt," says: "While Sir R. Abercrombie reconnoitred, the army continued under the most terrible and destructive fire from the enemy's guns to which troops were ever exposed. The work of death was never more quick, nor were greater opportunities ever afforded for destruction. Aim was unnecessary, the bullets could but do their office and plunge into the lines."

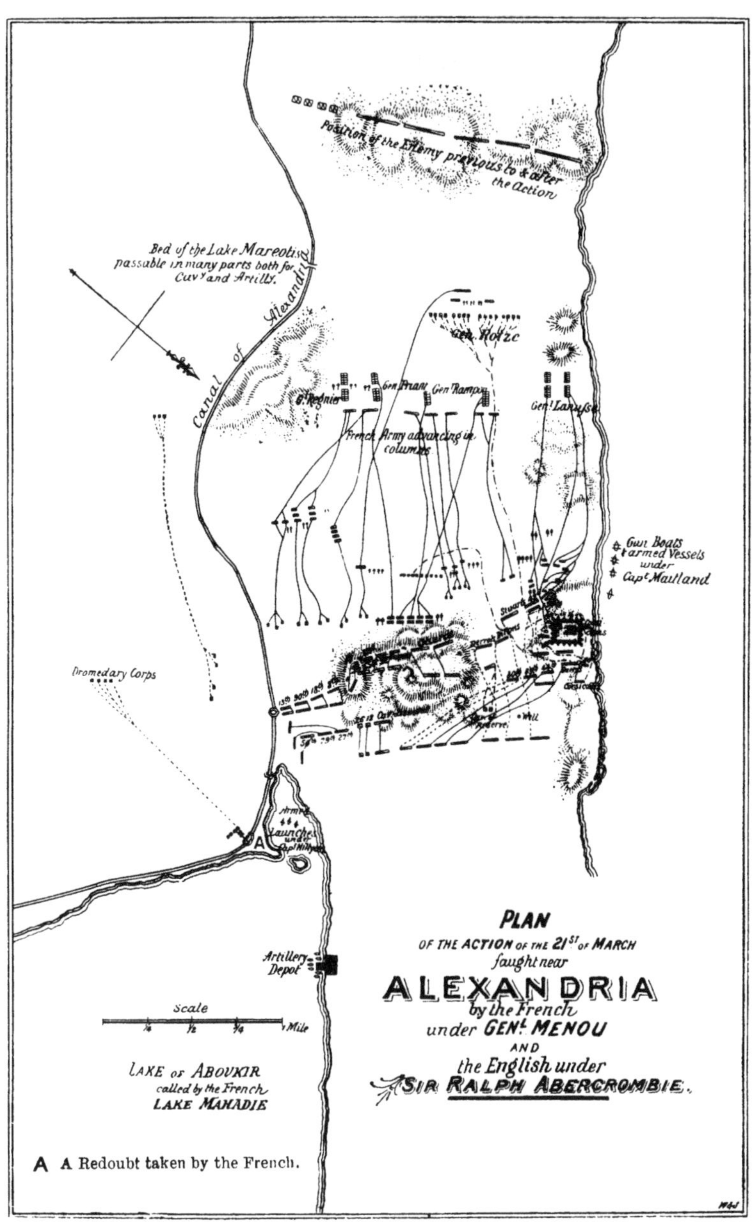
Position of the Enemy previous to & after the Action
Bed of the Lake Mareotis passable in many parts both for Cavy and Artilly.
Canal of Alexandria
Gen. Rampon
French Army advancing in columns
Gun Boats & armed Vessels under Capt. Maitland
Dromedary Corps
Artillery Depot
Scale
Mile
LAKE of ABOUKIR called by the French LAKE MAHADIE
PLAN
OF THE ACTION OF THE 21ST OF MARCH
fought near
ALEXANDRIA
by the French
under GENL. MENOU
AND
the English under
SIR RALPH ABERCROMBIE.
A A Redoubt taken by the French.

large ruins of an ancient palace, about 50 yards from the sea. This was held by the 28th and 58th Regiments, supported by the 23rd, 42nd, 40th, and Corsican Rangers. In the intervening flat between these heights and the right of the centre, were the cavalry of reserve; then the Guards on a hill, on their left. Forming an echelon with the Guards were the Royals, 92nd, and both battalions of the 54th; then, also in echelon with these, were the 8th, 18th, 90th, and 13th, and at right angles, with their left thrown back to protect the canal, and facing the lake, were the 27th, 79th, and 50th.

The remaining troops formed a second line. Four cutters were stationed on the right within 150 yards of the shore.

The labours of the army at this period were most arduous. Guns had to be brought up, batteries constructed, and all the ammunition and provisions brought from the magazine a mile and a half off, through heavy sand.

About this time tents were served out.

On the 20th of March a column of the enemy was perceived to enter Alexandria from the direction of Lake Mareotis, and information was received from an Arab chief that General Menou had come into that town with a large army, and that it was his intention to attack the British the following morning.

The British position had been strengthened by a battery erected to the left front of the ancient palace on our right. This palace was in ruins, and the openings into it had not been filled up. On the right front of the Guards' position was a redoubt, and a battery on their left, afterwards called the Citadel. There was a redoubt on the left of our line and two works on the Alexandria Canal.

The army was under arms as usual at 3 a.m. on the 21st of March. All was quiet until half-past 3, when a musket-shot was heard on the left, followed by the report of a cannon and scattered musketry.

It was evident that an attack was imminent, and through the dim haze of the early morning its development was anxiously awaited.

The left of our position on the Alexandria Canal, held by Lord Cavan's and General Craddock's brigades, where the first alarm had been given, was the weakest part of the whole line. There the French Dromedary Corps had seized an isolated redoubt (armed only with one gun) on the other side of Lake Mareotis and the Canal, killing or capturing all the defenders; but it soon became evident, from the smallness of the force employed, that no serious attack was intended there, and, beyond the formation of Lord Cavan's brigade on a new line almost facing the canal, no further notice was taken of it.

The real attack first developed itself on the right, where in the uncertain light the enemy had crept up close to our vedettes unperceived, and, following them up, vigorously attacked the ancient palace; but they were received by the 58th with so heavy a fire that they were forced to retire. Not to be denied, however, they were joined by another column and again advanced against the redoubt on the left face, defended by the 28th, while a third column forced their way in behind the redoubt, and even penetrated through the openings into the ruins of the palace. At this moment the 28th and 58th Regiments presented the extraordinary spectacle of troops successfully defending themselves against an attack on front, flanks, and rear.

The 23rd Regiment now came to the assistance of the 58th, and the 42nd to that of the 28th. As the 42nd approached the redoubt they were vigorously charged and broken by the enemy's cavalry, but though broken they retreated fighting, mixed up with the enemy, until relieved by the 40th Regiment.

It was here that Sir Ralph Abercrombie received the wound from which he died on the 28th, though he retained the command to the end.

Simultaneously with this attack, the French had advanced against the position held by the Guards on the right centre. Observing their echelon formation the French general attempted to turn their left, but General Coote at once advanced his brigade and the attack was effectually repulsed.

Beaten at all points the enemy retired about

10 a.m.; fortunately for them nearly all the British artillery ammunition and most of the musketry (the 28th at one time had to throw stones) was exhausted. As it was, they left behind them a standard, two guns, and 1,700 killed and wounded, of whom 1,040 were buried on the battlefield in the course of the next two days. It is calculated that, including prisoners, their loss must have amounted to nearly 4,000, amongst whom were most of their principal officers.

The loss of the English was 6 officers and 233 men killed, 60 officers and 1,190 men wounded, and 3 officers and 29 men missing.* The English tents were torn to pieces by the shot, and thousands of brass cannon balls were left glistening in the sand.

The 50th Regiment had 4 officers wounded, viz., Captain Ogilvy, Lieutenants Campbell and Tilsley, and Ensign Rowe; 1 private killed, and 2 sergeants and 35 privates wounded.†

* General Moore, though wounded in the leg early in the action, continued to command till the end, and Brigadier-General Oakes followed his example.

† The amount of wounded of the 50th Regiment seems out of proportion to the share they took in the action, the left never having been seriously attacked. They were, however, on the extreme left of the whole line, and probably provided the garrison of the isolated fort beyond the canal which was taken by the French at the commencement of the action. They were, at any rate, the nearest troops to it all through the action and the most exposed to its fire.

This fort was taken by the French Corps of Dromedaries (278 men) and 30 cavalry.

Lieutenant-General Lord Hutchinson now assumed command.

The army though victorious was numerically inferior to the garrison at Alexandria ; the fleet which assisted them was obliged to proceed to sea in rough weather, and the amount of ground included in the British lines was too extensive for efficient defence; it was, therefore, reluctantly decided to cut the canal of Alexandria, which would inundate all the low-lying lands to the south of Alexandria and partly isolate that city.

On the 13th of April this work was successfully completed, and the immense body of water continued flowing in for a month.

Already, on the 7th of April, Rosetta* (which had been attacked by a force composed of the 58th Regiment, 40th flank companies, 30 men of Hompesch's Hussars, with 8 cannon and 4,000 Turkish soldiers,† under Colonel Spencer) had fallen, and Fort Julien, between that place and the sea, and El Hamid, on the Nile to the south of it, had been besieged.

The success of the inundation enabled Lord Hutchinson to weaken his force before Alexandria, and on the 13th of April the 18th, 90th, and 79th Regiments, and a detachment of cavalry, marched to join Colonel Spencer, and were followed, on the 17th,

* Lieut.-Colonel Wauchope, 50th Regiment, was killed here.

† A Turkish force of 6,000 men joined the English on the 26th of March.

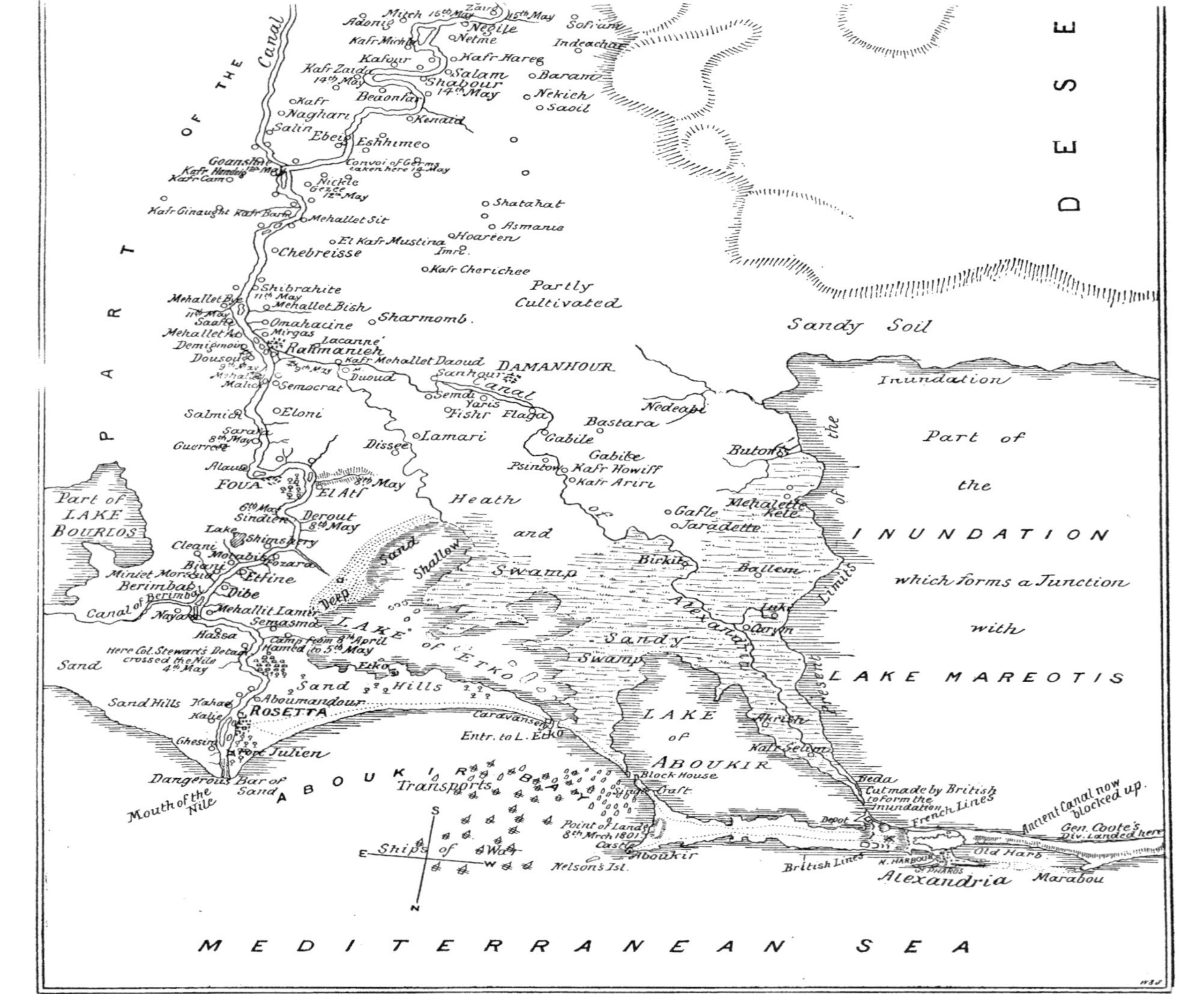
MEDITERRANEAN SEA
PART OF THE Canal
DESE
ABOUKIR BAY
Transports
Ships of War
LAKE of ETKO
LAKE of ABOUKIR
Part of the INUNDATION which forms a Junction with LAKE MAREOTIS
Part of LAKE BOURLOS
Partly Cultivated
Sandy Soil
Heath and Swamp
Sandy Swamp
Sand Hills
Deep Sand Shallow
Present Limits of the Inundation
ROSETTA
DAMANHOUR
Rahmanieh
FOUA
Alexandria
Aboukir
Nelson's Isl.
Point of Land 8th Mrch 1801
Mouth of the Nile
Dangerous Bar of Sand
Fort Julien
Caravanserai
Entr. to L. Etko
Block House
Cut made by British to form the Inundation
French Lines
British Lines
Ancient Canal now blocked up
Gen. Coote's Div. landed here
Old Harb
Marabou
Here Col. Stewart's Detat crossed the Nile 4th May
Camp from 8th April Hamed to 5th May
Canal of Berimbal
Alexandria Canal

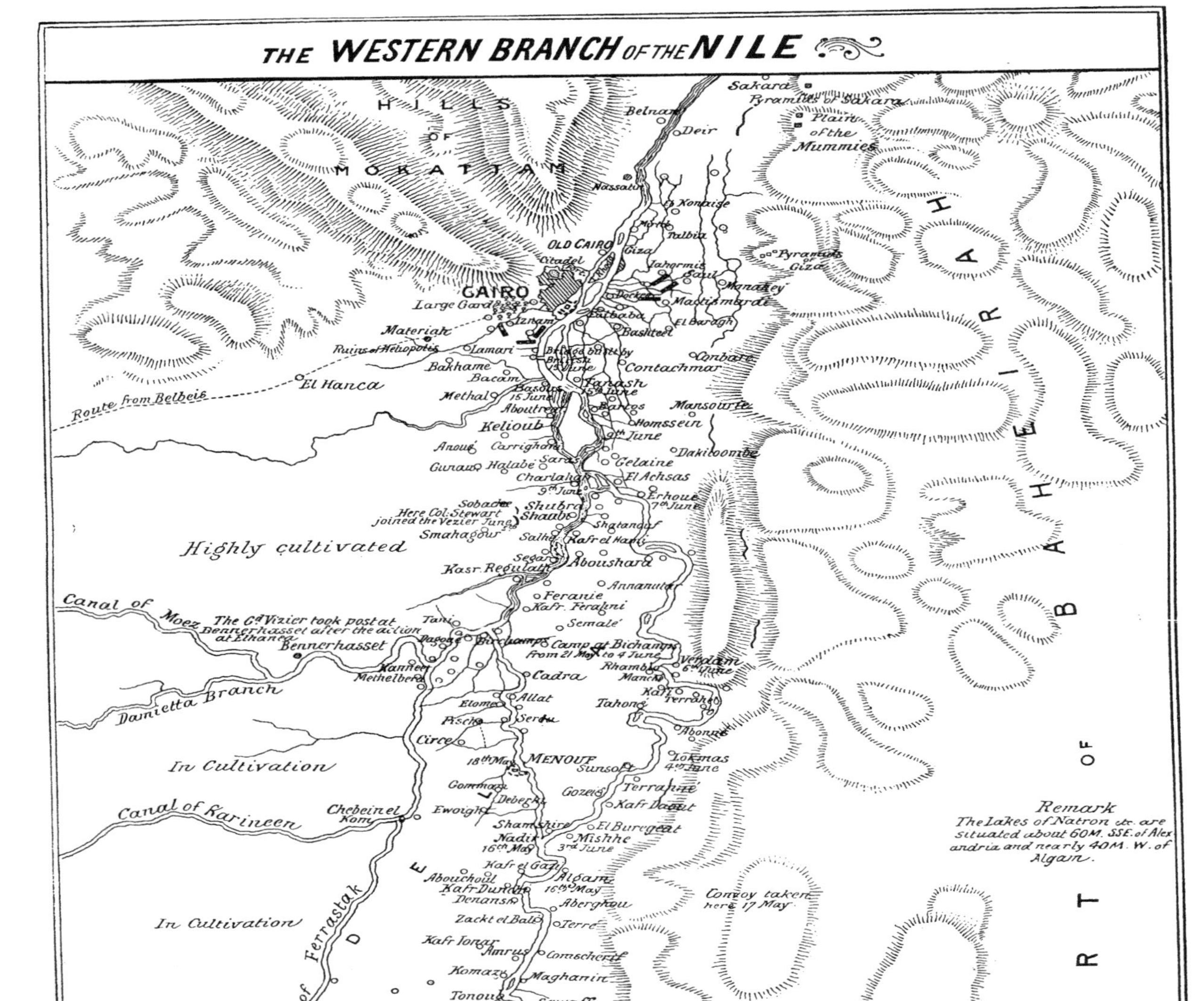
THE WESTERN BRANCH OF THE NILE
HILLS OF MOKATTAM
CAIRO
OLD CAIRO
Citadel
Large Gardens
Materiah
Ruins of Heliopolis
Lamari
Bakhame
Bacam
El Hanca
Route from Belbeis
Methal
Basous 15 June
Aboutreg
Kelioub
Anoué
Carrighan
Gurauo
Halabe
Saras
Charlaka
9th June
Sobacie
Here Col. Stewart joined the Vezier June
Shubra
Shaabe
Smahagour
Salha
Kafr el Hani
Highly cultivated
Segar
Kasr Regulath
Aboushard
Feranie
Kafr Ferahni
Semale
Annanutar
Canal of Moez
The Gd Vizier took post at Bennerhasset after the action at Elhanca
Bennerhasset
Tani
Dagoue
Bichamps
Camp at Bichamps from 21 May to 4 June
Manneen
Methelbera
Cadra
Rhamble
Manchi
Verdam 6th June
Damietta Branch
Etome
Allat
Tahone
Kafr Terrahie
Pische
Serdu
Circe
Abonne
In Cultivation
18th May
MENOUF
Sunsoff
Lokmas 4th June
Gommas
Debergh
Gozeis
Terrahie
Kafr Daout
Canal of Karineen
Chebein el Kom
Ewoight
Shamshire
Nadir 16th May
El Buregeat
Mishhe 3rd June
Kafr el Gaz
Algam 16th May
Abouchoul
Kafr Dunan
Denansh
Aberghau
Convoy taken here 17 May
Zackt el Bal
Terre
Kafr Ionar
Amrus
Comscherif
Komaze
Maghanin
Tonoue
Sowaff
Ferrastak
In Cultivation
Belnan
Deir
Sakara
Pyramids of Sakara
Plain of the Mummies
Nassala
Es Konaise
Monte
Talbia
Giza
Pyramids of Giza
Jahormie
Baul
Manakey
Dockes
Mattakmarde
Embaba
El Baragh
Basiteel
Conbare
Contachmar
Tarash 15th June
Barcos
Mansourie
Homssein
9th June
Dakiloombe
Gelaine
El Achsas
Erhoue 7th June
Shatanouf
BAHEIRAH
DELTA
RT OF
Remark
The Lakes of Natron &c. are situated about 60M. SSE. of Alexandria and nearly 40M. W. of Algam.

by the 30th and 89th, Generals Craddock and Doyle being appointed to the command.

Fort St. Julien surrendered on the 19th, after a brave defence.* This secured the command of the Nile.

On the 5th of May the main army marched in two columns.

The advanced guard consisted of Colonel Spencer's brigade.

New brigades as follows were now formed:—

Corps	Commander
The 11th Light Dragoons. The Corsican Rangers. 40th Flank Companies. The Queen's. 58th Regiment.	Colonel Spencer.
8th Regiment. 18th Regiment. 79th Regiment. 90th Regiment. 12th Dragoons and Detachment of 26th	General Craddock.
1st Regiment. 50th Regiment. 92nd Regiment. 30th Regiment.	General Doyle.

The Turkish army 4,000 strong, under Caia Bey, with 12 field pieces and 8 Turkish guns, also several

* Aboukir Castle had surrendered on the 18th. The garrison of Fort St. Julien consisted of 268 men, of which all except 160 were invalids.

Turkish gun-vessels and English armed and commisariat dgerms,* under Captain Stevenson, R.N., accompanied this force up the Nile.

General Coote with the remaining troops was left in command of the army before Alexandria.

A column consisting of the 89th Regiment, 20th Dragoons, and a body of Arnauts, amounting in all to about 1,200 men under Colonel Stuart, had been ordered to cross the Nile below Berimbal on the 4th, and to conform to the movements of the two columns of the main army; one of which marched along the Nile, the other by the shore of Lake Etko. The army halted on the 6th May in rear of Deroute, and Colonel Stuart's column advanced to a position between Sindien and Foua. Some Turks were sent forward to occupy the latter town, which the French evacuated, escaping to El Aft on the other side of the Nile.

On the 7th El Aft was abandoned, Colonel Stuart's position completely commanding the enemy's rear; and out of four French gunboats which attempted to escape, one only succeeded in doing so; two being sunk, and one being blown up. Here was found a paper containing authentic information of the number of the French infantry opposing the British advance, namely, 3,331; and here also,

* A dgerm must have been the equivalent of the dahabeiahs that are used on the Nile at the present day.

the cavalry reinforcement, which had been promised by the Grand Vizier, joined. Two thousand had been promised, but only 600 joined, and these were half-naked, wretchedly mounted, and badly disciplined.

On the morning of the 9th the army marched towards Rhamanieh, Stuart's column moving to Dousoug with the 89th leading.* He was attacked by a column of about 300 of the enemy, and waited for the co-operation of the gunboats, detained by absence of wind. He soon, however, continued to advance, the 89th, with great gallantry, lining the bank of the Nile in spite of heavy discharges of grape, and thereby preventing the escape of over 70 French dgerms, which attempted to pass but were forced back. Finally, this column took up a position opposite the French fort, the 89th in the centre, and the Turks on each flank.

About 4 o'clock in the afternoon Lord Hutchinson, thinking that the enemy might attempt to retreat towards Alexandria, moved the army forward to the Alexandria Canal; the Turks moving forward with their left on the Nile. The English infantry line, with their left refused, so as to align on the Turkish formation, consisted of General Doyle's brigade (of which the 50th occupied the left centre) on the left, General Craddock's in the centre, and Colonel

* In this and the movement up the Nile the sheiks greatly assisted the advanced column with information and supplies.

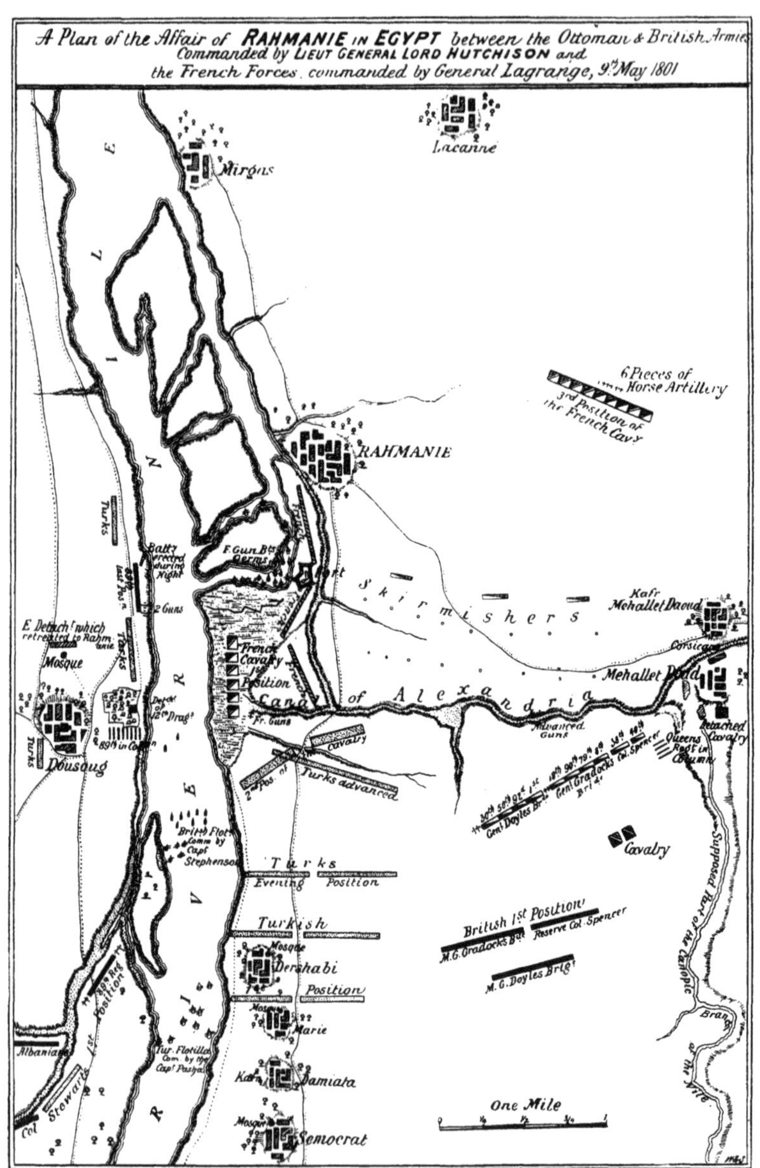
A Plan of the Affair of RAHMANIE IN EGYPT between the Ottoman & British Armies Commanded by LIEUT GENERAL LORD HUTCHISON and the French Forces, commanded by General Lagrange, 9th May 1801
Lacanne
Mirons
RAHMANIE
6 Pieces of Horse Artillery
3rd Position of the French Cavy
Skirmishers
Kafr Mehallet Daoud
Mehallet Daoud
Canal of Alexandria
French Cavalry 1st Position
Fr. Guns
Cavalry
2nd Pos. of Turks advanced
Advanced Guns
Queens Regt in Column
Detached Cavalry
Gen. Doyles Br.
Gen. Cradocks Br.
Col. Spencer
Cavalry
Supposed Part of the Canopic Branch of the Nile
British 1st Position
M.G. Cradocks Brt
Reserve Col. Spencer
M.G. Doyles Brig
One Mile
Turks Evening Position
Turkish Position
Mosque
Dershabi
Marie
Damiata
Semocrat
Tur. Flotilla Com. by the Capt Pasha
British Flot. Comm. by Capt Stephenson
Albanians
Col Stewart 1st Position
89th in Col.
12th Drag.
Dousoug
Mosque
E. Detacht which retreated to Rahmanie
Turks
2 Guns
F. Gun Batt.
Fort
NILE
RIVER

Spencer's on the right, with his flank regiment, the Queen's, in column in rear of the right.

The French cavalry took up a position to contest the passage of the canal.

General Lord Hutchinson, now informed of the position of Colonel Stuart's column on the reverse of the French forts, ordered a battery to be constructed during the night that would render this work untenable, and awaited the result in the position he had taken up.

His troops were under arms before daylight on the 10th, and as soon as day dawned they moved to occupy the positions appointed for attack. But before anything had been done, an officer came over from Colonel Stuart's column, to say that the fort at Rhamanieh had surrendered.

At 3 p.m. that day the French garrison of 110 men marched out; the remainder of their force having retreated towards Cairo.

It was noted that the plague was in the village of Rhamanieh, and all were forbidden to enter it.

The loss of the British in this affair was 4 officers wounded, 6 men killed, and 19 wounded. The Turks lost a larger number, and the French about 100. Their number had been augmented from Alexandria; and it was ascertained that their force consisted of 4,000 infantry and 800 cavalry, with 33 guns.

The difficult question now arose, whether it would be better to advance against Cairo, or to retire and besiege Alexandria.

There were great difficulties in the way of the first proposition; the men were already suffering from the plague, dysentery, and opthalmia, the commissariat difficulties would be great, and only salt meat could be obtained, the heat was intense, and the shoes of the men were worn out.

But, on the other hand, a Turkish army under the Grand Vizier was marching from Damietta to co-operate with us on Cairo, and a force of Indian troops under General Baird was expected from Suez. Not to advance might have caused these forces to be defeated in detail; while, without their assistance, the siege of Alexandria would be difficult and tedious.

It was, therefore, decided to march on Cairo. On the morning of the 11th May the army advanced; the entrenched camp at Rhamanieh being handed over to a garrison of 300 Turks.

The captured dgerms were employed in the transport of stores, and in carrying the men's knapsacks, for up to this point officers and men had carried everything themselves.

The army reached Kaffa Haudeig on the morning of the 12th. A sirocco was blowing, which parched the troops almost to suffocation, and the dgerms could not proceed. A short halt had therefore to be made. The British force reached Shabour, Sowaff, and Algam on the 14th, 15th, and 16th respectively.

On the 14th a valuable convoy, which had come

down the canal of Menouf, was captured; and on the 17th a still more important convoy was taken, escorted by 569 men—infantry, cavalry, and artillery, including 120 of the dromedary corps, one four-pounder, and 550 camels. These latter were most valuable. That evening the escort of the convoy embarked for Rosetta, *en route* for France.

Intelligence was received this day that the army of the Grand Vizier had defeated the French on the 16th at El Hanca; but, although defeated with a loss of 300 killed and wounded, they were enabled to retreat in good order on Cairo.*

The Turkish army now moved to Bennerhasset on the Damietta branch of the Nile, which brought it in touch with Lord Hutchinson's command.

On the 21st of May Colonel Stuart's column marched from Menouf to Birchamps; and at the latter village an interview took place between the Grand Vizier and Lord Hutchinson.

On the 1st of June the main army arrived at Mishlei; and officers of the artillery and engineers were sent back to Rosetta, to expedite the heavy artillery, &c., required for the siege of Cairo. Here

* The French force, which was estimated at 4,600 infantry 900 cavalry, and 24 guns, appears to have been the garrison of Rhamanieh retreating to Cairo.

As the Turkish army under the Grand Vizier amounted to 15,000 men, it ought not to have been difficult to cut the French force off from Cairo, when being between two armies and without supplies it must have surrendered.

the leader of the Mamelukes visited Lord Hutchinson, and assured him of the adhesion of his following.

The army now steadily advanced; the English army on the west bank, and the Grand Vizier's army, which Colonel Stuart's column had now joined, on the east bank of the Nile; and on the 15th June we find the former at Tanash, and the latter at Bassous.

On this date, Lord Hutchinson sent a summons to surrender to the commandant at Cairo, who refused to enter into any negotiations.

On the 16th* a bridge of boats was commenced over the Nile at Shuha, to complete the communication between the two armies.

On the 19th, the Vizier's army moved forward within gunshot of Cairo, and the first works against that town were established between Elwoine and Elmini. In rear of the latter, Colonel Stuart's brigade was afterwards posted.

The English army advanced to within a mile and a half of the Giza suburb of Cairo on the 21st, encamping in two lines; and the Mamelukes drove in a cavalry piquet, and occupied the village of Sachatmichle, within 300 yards of the works of that place.

* Colonel Wilson's expedition to Egypt, says on this date: "A sirocco wind darkened the atmosphere with a burning mist, the thermometer was 120° in the shade, and the ground was heated like the floor of a furnace." He also says, "On the 16th the garrison of Cairo fired a *feu de joie* from all their guus, which afterwards turned out to be in honour of 'the capture of Ireland.'"

Night and day, the greatest exertions were now made to drag up the heavy cannon, and the requisite ammunition.

The 28th and 42nd Regiments arrived at this period from Rosetta under General Hope. General Moore and Brigadier-General Oakes (who had recovered from their wounds) also accompanied these troops. *

On the 22nd of June, a French officer came in with a flag of truce asking for a conference; and, in consequence, General Hope met the French General Moran under the trees near Giza. The French general stating that he was instructed to negotiate for an evacuation of Cairo, a committee was formed on both sides to arrange details. This committee met on the 23rd and following days. The articles of capitulation were agreed to, on the 26th.

On the 23rd a cessation of hostilities was proclaimed, and on the 27th the articles of capitulation were signed. (See Appendix for articles of capitulation.)

The French evacuated Cairo during the night of the 10th June, and Colonel Stuart at once occupied the citadel with the 89th Regiment.

At daybreak on the 15th, the French totally evacuated Giza, and, with the allied army, began their march for Rosetta in the following order:

The Turkish army preceded, the British followed,

* See note on page 75.

then the French army with flanking parties of their own cavalry on their left. The English cavalry and the Mamelukes brought up the rear.

The command of the army devolved on General Moore, General Hutchinson remaining behind, to settle arrangements for the government of Egypt.

General Craddock was too unwell to proceed with the army.

The Indian and Cape contingents, parts of which had arrived at Cossir, on the Red Sea, in June, were ordered to march across the Desert to Cairo. General Moore's force proceeded uneventfully to Rosetta, the French forming every night three sides of a square (the Nile being the fourth), with their artillery and baggage in the centre.

At Déroute the French army took the lead in order to be in readiness for embarkation. The whole force encamped at El Hamed, four miles from Rosetta on the 28th; the next day Lord Hutchinson arrived from Cairo, and on the 31st the first division of the French army embarked at the caravansary (at the entrance to Lake Etko) for France.* But the whole embarkation was not completed till the 7th July.

General Coote's division had remained before Alexandria, having been strengthened early in July with reinforcements from England, consisting of the

* The French embarkation returns show a strength of 13,754 men, including civilians and auxiliaries, but exclusive of women and children.

22nd Dragoons, a detatchment of Guards, the 20th, 24th, 25th, and 26th Regiments, the Irish Fencibles, the foreign levies of Waterville (Swiss), and the Chasseurs Britannique, with drafts for several regiments.

The only event of importance that had occurred, was a further cutting of the Alexandria Canal by the French on the 20th June, so as to extend the inundation, and contract the front for siege operations against Alexandria. This General Coote met by constructing a dam parallel with the canal 150 yards long, the extremities of which rested on the higher ground. This dam could only be constructed during the night, and took nearly a month to complete.

General Doyle's brigade from Cairo (which included the 50th Regiment) marched into the camp before Alexandria on the 11th July, and was followed by General Moore with the Reserve on the 13th. On that date English and Turkish gunboats forced their way, through the inundation in Lake Aboukir on to Lake Mareotis, compelling the French boats to take refuge near the shore.

General Hutchinson arrived on the 15th, and preparations were at once made to besiege also the western side of Alexandria.

With this object, boats provisioned for three days were assembled on the inundation to the left of our position, and a division was embarked under Major-General Coote on the evening of the 16th, composed of the Guards, the 25th, both battalions of the 27th, the 44th, the 26th, two battalions of the 54th, and 100 of the 26th Dragoons—in all about 4,000 men. These sailed during the night.

The 50th, the 92nd, and the 30th Regiments were ordered to make a diversion on the eastern side at daybreak next morning, with which General Moore was to co-operate.

General Doyle, who had been ill at Rosetta, at once returned, and took over the command of his brigade, which was ordered to form the left column, and attack the Green Hill on the right of the French position. The 30th Regiment was to move against the advanced work on the left, the 50th against that on the right, and the 79th were to remain in reserve at the foot of the hill.

General Moore, with the right column, was at the same time to attack the Nole Hill on the right of the enemy's position.

Both columns took up their respective posts with very little opposition, and General Moore having reconnoitred the enemy's position from the Nole Hill, found that it was untenable, and withdrew his column.

As soon as the French found their piquets attacked, they expected an assault on Alexandria and beat to arms, commencing a heavy fire, which was continued without intermission for three hours, although doing little damage.

About seven o'clock, a body of some 600 French rapidly advanced, against that part of the Green Hill held by the 30th Regiment. The men of that regiment had been ordered to shelter themselves, from the heavy fire in the ditches of the works, and in the inequalities of the ground. They were thus scattered, when the French began the ascent of the Green Hill, supported by a heavy fire of shell, round shot, and grape, from all their batteries.

The 30th Regiment, directly they became aware of this attack, sounded the assembly. Then 170 rank and file were collected, and, as the French column had nearly gained the top, Colonel Lockhart,* commanding the regiment, without waiting for orders, at once ordered a charge, completely routing the enemy, who left 10 prisoners and 100 killed and wounded behind.

Directly General Doyle perceived this movement of the enemy, he ordered the advance of the 50th and 92nd regiments, but the distance they had to pass over, was too great for them to render much assistance.

In the meantime, General Coote's division had sailed towards Marabou, and, finding the promontory of land near there occupied by the French, he landed about three miles further on.

* Colonel Spencer who had been in command of the brigade during General Doyle's absence, was present with the 30th Regiment.

That evening the French, seeing the allied gunboats prepared to attack them, abandoned their own gunboats, which they attempted to blow up. Five of them, however, were saved and captured.

During the night of the 17th, trenches were opened against the Fort of Marabou, on an islet separated from the main land by a reef of rocks 150 yards long, fordable except in the centre.

It consisted of a regular fort, with a tower in the centre.

About 12 o'clock in the morning of the 20th the tower of Marabou fell, and a storming party was told off to attack the fort; but a summons to surrender was first sent, which, after some delay, was agreed to.

General Coote advanced his division towards Alexandria at 6 a.m. on the 22nd. He divided his army into three lines, marching in three columns of half brigades.

The right column moved along the flat between the lake and the range of hills, on which a French force under General Eppler was posted; the centre was directed on the hills, and the left column on the flat near the sea. Two hundred of the Guards formed the advanced guard, and the Dragoons, with six guns, marched in rear of the left. Captain Stevenson, with the allied gunboats, moved on the right.

The French opened a heavy fire from all their guns, the British pieces replied, and the army continuing to advance, the French retreated to the ridge in rear, abandoning a heavy gun.

Though galled by musketry and grape, General Coote still pressed on till within 1,400 yards of Alexandria, the French abandoning their tents and baggage in their hasty retreat.

On hearing of the success of General Coote, Lord Hutchinson ordered Colonel Spencer, with 1,500 men, to join him,* and, as the boats could not be got ready till night, Lord Hutchinson, fearing a sortie on General Coote's division, ordered men from General Moore's and General Craddock's brigades to make a false attack, by crawling up in the dark as near Alexandria as possible, and firing into the works. This ruse was completely successful; the French beat to arms, and kept up a heavy fire till daylight, with a resultant loss to the English of only one man killed and one wounded.

Two batteries were opened on the redoubt "Des Bains" (western side) on the 25th, and that evening the 20th Regiment, with a detachment of dragoons, gallantly cut off the French piquet in front of that work.

On the morning of the 26th, the English batteries at the eastward side (on the Green Hill), opened against the right of the French position.

That evening General Menou sent his first aide-de-

* Colonel Spencer was second in command of General Doyle's division.

camp to ask for an armistice for three days, in order to arrange terms of capitulation. This was granted; but at the expiration of the three days a further extension for thirty-six hours was asked for; this being declined, an extension till 2 o'clock the next afternoon was asked for and granted. At the expiration of this time an aide-de-camp brought the articles, many of which were refused, and at 11 o'clock that night the aide-de-camp returned with the articles agreed to, as corrected by General Hutchinson.

The next day, August 31st, General Hope went into Alexandria and signed the articles of capitulation, which Lord Keith also signed on the part of the Navy, on 2nd of September. (See Appendix.)

The total number that surrendered, including civilians, amounted to 11,213.

At 11 o'clock on the 3rd of September, the grenadiers of the army in three columns, with drums beating and colours flying, took possession of Alexandria.

General Baird, with the Indian army, had now arrived from Cairo, and was encamped at Aboumandour, near Rosetta.

The thanks of Parliament were voted to those gallant and meritorious officers, who had so ably and eminently distinguished themselves throughout the war.

The expedition to Egypt having been happily terminated, the embarkation of a division of the army, under General Craddock, was decided on.

The troops ordered to be embarked were the 1st 18th, 30th, 44th, 50th, and 89th, amounting to 5,031 men. The first division of these, consisting of the 1st, 30th, and 89th, embarked at Aboukir on the 10th September, and sailed on the 12th. On the 15th of that month, however, Lord William Bentinck arrived with despatches from England, which altered the destination of this force.

The 50th Regiment, therefore, proceeded to Malta on the 17th of October.

Here it received drafts from various regiments, and Lieutenant-Colonel Walker took over the command.

The regiment returned to Ireland, landing in Cork on the 4th May, 1802.*

A second battalion was formed at Colchester on the 1st of October, 1804, under Lieutenant-Colonel Rowe.

New colours were received by the regiment at the end of 1804, bearing for the first time "the Sphynx" and the word "Egypt" on them.

The old colours were then brought out and burnt, in front of the first battalion, with military honours.

Copenhagen.

The Emperor Napoleon having made himself master

* As the 50th Regiment are said to have obtained the nick-name of the "blind half-hundred," from the amount of men who suffered from opthalmia in this campaign, it will be interesting to note that on the capitulation of Cairo, a return gives the number of men of the English army who had totally lost their sight from this disease on that date as 160 while 200 had lost one eye.

of the continent, a design was attributed to him of massing all the European navies, for the subjugation of England.

Under the circumstances the British Government took the bold resolution of securing possession of the Danish fleet until the conclusion of the war. It was hoped at first that this might have been accomplished by diplomacy, but it was eventually found necessary to besiege Copenhagen and capture the fleet.

The sea defences of Copenhagen consisted at this time of a battery built on piles at the entrance of the canal, an arsenal and harbour, mounting 68 guns, besides mortars; another pile battery in front of the citadel, mounting 86 guns and 9 mortars; and the citadel, which mounted 20 guns and 12 mortars. There were also block ships, floating batteries, and from 25 to 30 gunboats, all ready for action, whilst in the arsenal lay a fleet of 16 sail of the line and 21 frigates and sloops, besides three 74's on the stocks, one being nearly ready for launching. (Cust's "Annual of the War.")

The 1st battalion of the 50th Regiment embarked at the Cove of Cork for Ramsgate on the 5th of June, 1807. The regiment proceeded from Ramsgate to Deal, where all the effective men of the 2nd battalion were drafted into it.

It re-embarked on the 25th July under the command of Lieutenant-Colonel Walker, and sailed with the expedition under Lord Cathcart for Copenhagen, landing at Charlotte Land (near the capital) in the island of Zealand on the 16th August, and marching by the coast to Carlottenburg.

The 50th was brigaded with the 32nd and 82nd Regiments under Major-General Spencer, and formed part of the 2nd brigade of the 2nd division, under Lieutenant-General Sir David Baird.

At daybreak on the 17th of August, the three British columns marched by their right, to invest the town and fortress of Copenhagen.

Major-General Spencer's brigade being on the left of the British line,* was attacked by the enemy about

* Lord Cathcart writes: "Lieutenant-Colonel Smith with the 82nd held the post at the Windmill, the most exposed to fire from the gunboats and to sorties."

It is probable that the 82nd as the junior regiment of the brigade occupied the centre, and the 50th Regiment as the second senior would have been on the left, and as Major-General Spencer's brigade was on the left of the British line. The 50th Regiment would thus have been on the extreme left of the whole line.

The following is a return of the strength of the Regiment at Copenhagen (War Office Returns):

Lieutenant-Colonel	1	Surgeon	1
Majors	2	Assistant Surgeons	2
Captains	6	Sergeants	48
Lieutenants	9	Drummers	21
Ensigns	6	Privates	840
Paymaster	1	Present } Sick	28
Adjutant	1	Absent } Sick	55
Quartermaster	1	On Command	29
		Total	952

noon on the above date, and the Danish gunboats cannonaded them with grape and round shot. The piquets however drove in and pursued the enemy, and afterwards resumed their posts.

Works at the Mill were commenced against the town on the 19th, and carried on by working parties of 600 men, relieved every 4 hours.

On this date also the foundry and depôts of cannon of Frederichswork were taken by surprise.

On the 20th a force of the enemy's cavalry and infantry, having been observed by the piquets of the left, they were driven in and pursued to the gates of Roeskild.

The trenches were pushed forward on the 21st, and a new battery erected 300 yards in advance. Lord Rosslyn's corps of the King's German Legion landed in Kioge Bay, and took part in the second line on that date.*

The army was under arms at 3 a.m. on the 24th,

The following officers of the 50th Regiment held posts on the staff:

Lieutenant-Colonel George Tucker, D. A. A. G.
Captain Henry Riddle, A. Q. M. G.

Captain Charles Platt ,, Thomas Snow Lieutenant John Kent ,, George Armstrong Gent. W. H. Pitkin	Acting Assistant Commissaries.

* Lord Rosslyn's Division consisted of eight battalions of the King's German Legion and one independent company, Major-General Von Drechsel being second in command.

when a general advance was made against the enemy's works. The British centre taking post on the heights near the town, the enemy's piquets were driven in, and the Brigade of Guards occupied the summits. Sir D. Baird's division, to which the 50th Regiment belonged, turned and carried a redoubt which the enemy had been constructing some time, and which was that night converted into a work against them. Part of the suburbs of the town were set on fire by the enemy.

The old works were now abandoned, and a new line of works commenced within 800 yards of the town, and still nearer on the flanks.

On the 26th the enemy's gunboats made an attack on the left of our position, but were driven back by our batteries at the Windmill. An unsuccessful sortie was also attempted.

The Danish army having now approached, Sir A. Wellesley's division attacked and defeated it on the 29th, taking many prisoners, cannon, and stores.

The enemy attempted a sortie before sunrise on the night of the 31st, our batteries being nearly completed, and two-thirds of the ordnance mounted.

They were promptly engaged and repulsed by the piquets of the 50th Regiment, commanded by Lieutenant Light; which, arousing all the other piquets, the enemy was compelled to retire with loss. General Baird, commanding the division, was twice wounded in the above, but did not quit the field. Lieutenant

Light was promoted to a company on the 14th for his conduct in this affair.

The mortar and other batteries having been completed, the place was surrounded on the 1st of September, and capitulated on the 7th.

The loss of the 50th Regiment to the 31st of August was, 1 private killed, Ensign Bilson and 15 privates wounded.

In October a brigade was formed, consisting of a light brigade of Artillery, and the 23rd, 50th, and 79th Regiments, under Major-General Spencer, to assist the King of Sweden; but he having declined the proposed aid, they returned with the rest of the army to England, and the 50th Regiment landed at Deal in November, 1807.

CHAPTER V.

VIMIERO.

THE first battalion of the 50th ("The Queen's Own") Regiment embarked under sealed orders at Portsmouth on the 17th of December, 1807, under the command of Major-General George Townsend Walker, upwards of 1,000 strong.

They formed part of a brigade consisting of the 29th, 32nd, 50th, and 82nd Regiments, under the command of Major-General Sir Brent Spencer. The transports conveying the regiments encountered so severe a storm in the Bay of Biscay that the fleet was scattered, some of the vessels turning back to Plymouth, and other ports; the vessels which carried the head-quarters taking refuge at Messina. The whole Regiment eventually assembled at Gibraltar.

The "Queen's Own" again embarked and sailed for Cadiz on the 13th of May, 1808. They were present with the blockading fleet outside the harbour when the Spaniards opened fire on the French fleet within, and compelled them to surrender. The regiment afterwards landed at Port St. Mary's, and remained on shore for a week. Re-embarking on the 22nd July, they sailed for Portugal and landed at Figueras at the mouth of the Mondego river.

The 50th were now brigaded with the 45th and 91st Regiments, under the command of General J. Catlin-Crawford.

The first skirmish with the enemy, under General La Borde, took place at Brilos, near Obidos, on the 15th of August. On the 17th the light company of the regiment, under Captain Harrison, was engaged with the enemy and behaved with great gallantry. The same day the British army, under the command of Sir Arthur Wellesley, advanced against the strong position which General La Borde had taken up at Rorica. On that occasion the central attack, under Sir Arthur Wellesley himself, was composed of Hill's, Nightingale's, and Fane's brigades, with J. Catlin-Crawford's brigade (which included the 50th Regiment) in reserve. The position was stubbornly defended; but though the enemy were eventually compelled to retreat, the victory cost us nearly 500 killed, taken, and wounded, out of about 4,000 men engaged.

Sir Arthur Wellesley now moved on Vimiero to cover the disembarkation of troops. Here a new commander-in-chief (Sir Harry Burrard), who forbade any offensive movement, having arrived, Sir Arthur Wellesley was compelled to remain at Vimiero.

Line of retreat from right attack
Position of the French when Sir A. Wellesley proposed to follow up the Victory
Solignac's situation when the order from Sir H. Burrard stopt Gl. Ferguson
Ferguson's position when his progress was arrested
Breniers attack
Solignac's attack
Portuguese
Ferguson
Nightingale
Bowes
Breniers
Laborde
Kellerman
Cavalry
Kellerman
Loison
Ackland
VIMIERO
Hill
Road to Torres Vedras
2 Regts. that arrived from Lisbon at the close of the attack
to Torres Vedras
R. Maceira
Original position of the English right
Porto Novo
THE SEA
Sketch of the
BATTLE OF VIMIERO
21st. August, 1808.
Scale
Miles
0 1/4 1/2 3/4 1 1/4 1/2 3/4 2
2 4 6 8 10 12 14 16
Furlongs

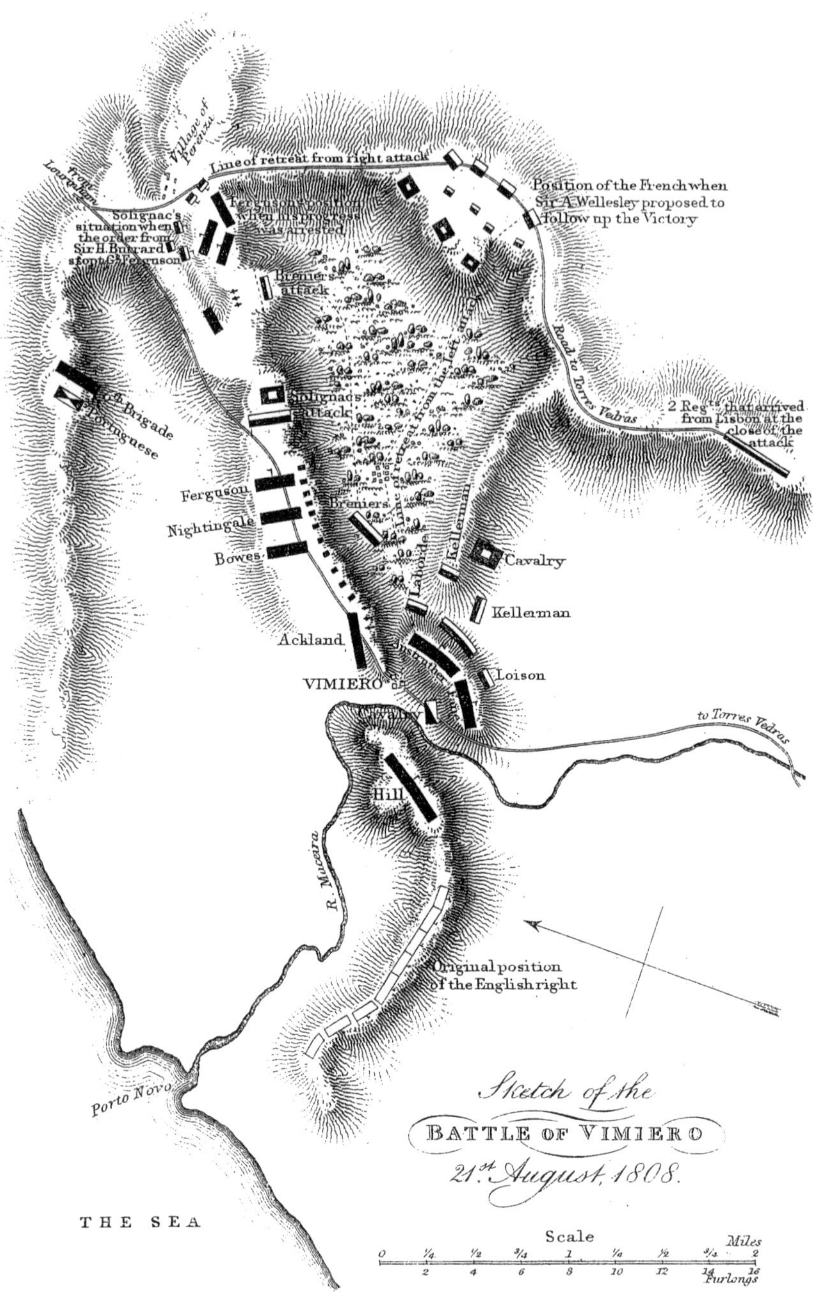

P. 101

The 43rd and 50th Regiments, and the 95th Rifle Corps, were formed into a light brigade, under the command of General Fane. The early morning of the 21st discovered the enemy, who had concentrated his troops under Marshal Junot, advancing to the attack. At 7 a.m. on that date the advanced guard of the French horse was seen to crown the heights to the southward, followed by a force of infantry preceded by other cavalry. Column succeeded column in order of battle, moving along the road from Torres Vedras to Lourinham.

At this juncture the British army occupied the following position:—1st, 2nd, 3rd, 4th, and 8th Brigades on our right on a mountain which, commencing on the coast, swept in a half-circle close behind the right of the Vimiero Hill, and partly commanded it. This was terminated on its left by a deep ravine dividing it from another strong and narrow range of heights, over which the road from Vimiero to Lourinham passed. In front of this ravine, forming, as it were, the apex of a triangle of which these two ranges formed the sides, was the Vimiero Hill. Some little distance behind this, and in the deep ravine above mentioned, stood the village of Vimiero on the little river Maceira. Here also were the parc and commissariat stores.

The right of this hill was held by General Fane's brigade, of which the 50th Regiment occupied the centre, the left being held by General Anstruther's brigade. General Fane's left rested on a churchyard, blocking the road which led over this height to Vimiero. The 5th Brigade, and Portuguese, were on the recurve towards the coast of the left range of hills.

As the advance of the French on the Torres Vedras road did not threaten our right, the 2nd, 3rd, 4th, and 8th Brigades* were moved across the ravine from the right to the left range of heights, leaving on the right range only the 1st Brigade under General Hill. This left General Fane's brigade on the extreme right, except General Hill's brigade, which was on the other side of the ravine and of the river Maceira, General Fane's brigade being considerably in advance.

The ground between the French and English armies was so wooded and broken, that after the French had passed the ridge where they had first been descried, no correct view of their movements could be obtained. They had 14,000 fighting men organized in four divisions. Generals La Borde and Brennier were directed against our left and centre; but, becoming entangled in the ravines, the disposition was imperfectly carried out; and another brigade, under General Solignac, was ordered to turn our left. General Kellerman moved with the reserve of grena-

* General Fane had under his command on this occasion the 60th Rifles, 4 companies of the 95th Regiment, and the 50th Regiment.

diers behind General Loisson to attack our right (the Vimiero Hill).

Generals Loisson and La Borde now formed one principal and two secondary attacks against the Vimiero Hill. One of these secondary attacks advanced against Anstruther's brigade. The other endeavoured to penetrate by the road, which passed between the ravine and the church on Fane's left, while a massive column under General Loisson,* consisting of five regiments in close order of half battalions, and preceded by artillery, dashed against our right, or that part of the Vimiero Hill which was now only held by the 50th Regiment, one company of the 95th† Regiment, and three guns under Lieut.-

* Sir William Napier, in his "History of the War in the Peninsula," says: "The main column under La Borde, preceded by a multitude of light troops, mounted the face of the hill with great fury and loud cries. The English skirmishers were forced back upon the lines; and the French masses reached the summit, but shattered by the terrible fire of Rode's artillery and breathless from their exertions. In this state, being first struck with musketry at the distance of half pistol shot, they were charged in front and flank by the 50th Regiment and overthrown."

The despatch of Colonel Walker, however, of October 17th, 1812, is so minute in the description of this column taken from an order of battle signed by Charlot, a brigadier of the 7th French Division, which was found on a French colonel who was killed in the action (and was afterwards confirmed to Colonel Walker by General Loisson himself), that I am compelled to adopt his statement that this column was commanded by General Loisson and not by General La Borde.

† The other companies of the 95th had been ordered away.

Colonel Robe, Royal Artillery. The 60th Rifles and three companies of the 95th, which had originally also defended this part of the hill, had been moved away prior to the attack.

The country was overspread with vineyards, interspersed with chestnut and olive trees, thick woods in the distance forming a background.

In advance of the regiment, and on lower ground, was a piquet under the command of Captain Thomas Snow, 50th Regiment. About 8 o'clock, the fire becoming heavy, this piquet was reinforced by two other companies of the regiment under Captain Coote. In this position the piquets were exposed to a very heavy fire, and Captain Coote having been shot through the heart, and Captain Snow having been detached to occupy some woods on the left, the command devolved on Lieutenant Mark Rudkin, who now gave orders to retire; and the piquets, extending to the right and left, fell back under a shower of bullets. Taking advantage of the shelter of olive-trees and vines, and alternately firing and retreating, they eventually gained a rising ground a little in advance of the rest of the battalion, where their excellent fire contributed in the sequel very materially to the success of the action, by attracting the attention of the enemy to that flank during the manœuvres performed on our right. In consequence of the absence of these three companies, the left wing of the regiment was necessarily extended with intervals. It occupied the most commanding ground to cover the passage to the town.

This was the position held by the 50th Regiment when General Loisson directed his main column, upwards of 5,000 strong, against that part of the hill held by the Regiment, which did not muster 900 men. General Anstruther's brigade on our left was at the time so fully occupied with General Kellerman's attack that it was unable to give any assistance.

General Loisson's attack is so ably described in the despatch* of Colonel G. T. Walker, commanding the regiment, of October 17th, 1812 (O.R.R.), that I cannot do better than give it in his words.

"A massive column of the enemy composed of five regiments in close order of half battalions, supported by seven pieces of cannon, and under the command of the general of division, Loisson, made a rapid march towards the hill, and though much shaken by the steady fire of the artillery, after a short pause behind a hedge to recover, it again continued to advance; till Lieutenant-Colonel Robe, R.A., no longer able to use the guns, considered them lost. Up to this time the 50th had remained at ordered arms, but as it was impossible, on the ground on

* I have had to make slight corrections, altering evident clerical errors, in order to make the meaning intelligible. It will be noted that the despatch, though signed by Colonel Walker, speaks of him in the third person. I tried to find the original despatch in the Record Office, but was unable to do so. Lieutenant-Colonel Walker was Brevet-Major-General.

which it stood, to contend against so superior a force, and Colonel Walker, having observed that the enemy's column inclined to the left, proposed to Brigadier-General Fane to attempt to turn its flank by a wheel of the right wing. Permission for this having been obtained, this wing was immediately thrown into echelon of companies of about four paces to the left, advanced thus for a short distance, and then ordered to form line to the left. The rapidity, however, of the enemy's advance, and their having already opened a confused though very hot fire from the flank of their column—though only two companies of the wings were yet formed—these were so nearly in contact with and bearing on the angle of the column that Colonel Walker, thinking no time was to be lost, ordered an immediate volley and charge. The result exceeded his most sanguine expectation. The angle was instantly broken, and the drivers of the three guns advanced in front, alarmed at the fire in their rear, cutting the traces of their horses, and rushing back with them, created great confusion, which by the time the three outer companies could arrive to take part in the charge, became general. Then this immense mass, so threatening in its appearance but a few minutes before, became in an instant an ungovernable mob, carrying off its officers and flying like a flock of sheep, almost without resistance, for upwards of two miles. On clearing a wood, Colonel Walker, observing a party of cavalry to be drawn up on a small plain threatening his flank, deemed it necessary to put a stop to the pursuit, as a party of the 20th Dragoons, which had previously joined in it, had already (through getting entangled in a wood) suffered so seriously as to be incapable of affording any further assistance. Having from hence reported his situation and received Brigadier-General Fane's orders, Colonel Walker retired with the regiment to his former position, while the enemy continued their retreat eastward in a direction different from that of their resources.

"The immediate result of this success of the 50th Regiment (including the assistance derived from the artillery during the advance of the enemy and from the 20th Light Dragoons in their retreat) was 1,000 killed, 360 prisoners, and 6 pieces of cannon, the Regiment not mustering at the time 900 men in the field, and the enemy's column being considerably above 5,000. General Anstruther's brigade on the right, had been too fully occupied with the attack of Kellerman's reserve, to be able to take any steps whatsoever, in the immediate support of the 50th Regiment. A copy of the order of battle, found on a French colonel killed in the action, countersigned by Charlot (one of the brigadiers of the division), and afterwards personally acknowledged by him, confirms this statement."

Meanwhile, General Anstruther had repulsed the attack made upon him, and had sent the 2nd Battalion of the 43rd Regiment to the churchyard, from which our left was at one moment threatened; for Kellerman having reinforced his attack on that point it was temporarily successful, but the 43rd rallying, with a hot fierce struggle, drove it back in confusion.

The French now fell back along the whole front; and Colonel Taylor, riding out with the 20th Light Dragoons, fell on their disordered masses; but following them too far, and becoming entangled among the vineyards, they were in turn charged by superior numbers of French cavalry and overthrown, with the loss of half their numbers and their colonel slain.

As far as the eye could reach over the thickly planted valley, and across the open country lying beyond the forest, the fugitives were running in wild disorder, their white sheepskin knapsacks discernible among the far distant woods.

The ground was thickly strewed with muskets, side arms, accoutrements, and well-filled knapsacks, which had been hastily flung away.

The loss of the 50th consisted of Captain Coote killed, Major Charles Hill and Captain J. N. Wilson wounded, and 2 sergeants, 2 corporals, 1 drummer, and 38 privates either killed or died of wounds. They captured a standard, pole, and box, which were borne by a sergeant between the colours during succeeding campaigns. (Patterson.)

The regiment bivouacked that night in the position they had occupied previous to their successful charge, where they remained the following day (the 22nd). On the 23rd August they commenced their march to Lisbon, negotiations for the convention of Cintra having stopped further hostilities. They were well received everywhere on their route through Portugal; and on arrival at Lisbon they were encamped on an elevated space, called Campo Santa Anna, in conjunction with the 29th, 40th, and 79th Regiments.

After the embarkation of the French in accordance with the convention of Cintra (August 30th, 1808), the 50th Regiment proceeded to Monte Santo,* where they arrived on the 28th September.

Colonel G. T. Walker having obtained leave of absence after the battle of Vimiero, the command of the regiment devolved on Major Charles Napier, who had joined from the second battalion to replace Major Hill, severely wounded.

After the above action fifteen volunteers from the French 70th Regiment joined the 50th, and their long red plumes were afterwards worn as trophies by the band of the regiment.

CORUNNA.

The 50th Regiment left Monte Santo in pouring rain, at 6 a.m. on the 28th October, 1808, and com-

* Monte Santo was about four miles from Lisbon, on the road to Cintra, which lay about fourteen miles north-west from Lisbon.

menced their march along the Tagus and across the Zezere river to join Sir John Moore at Salamanca. They marched viâ Abrantes along the right bank of the Tagus, and halting at Sacarem, arrived at Villa Franca, where some boots were issued from the stores there on the 29th of October. They reached Azambuja on the 30th and Santarem on the 31st, where they remained until November 3rd.

Passing through Cuidad Rodrigo they reached Salamanca on the 25th of November. The regiment was brigaded with the 4th and 42nd Regiments, under Lord William Bentinck, and formed part of Sir David Baird's division. They marched out of Salamanca with the army of Sir John Moore on the 12th of December, the snow lying deep on the ground, and a wintry blast coming down from the hills. They were closely followed by the armies under Marshal Soult, before whose vastly superior forces Sir John Moore was compelled to retreat, at first towards Vigo, afterwards towards Corunna.

The passage of the river Esla, near Valencia, on the 26th of December, was the first difficulty encountered, for there was a rapid current and a bad ford; and, in spite of the wintry weather, the men, holding their arms and ammunition above their heads, had to wade through the river up to their middles, and sometimes up to their necks. Nor were the women and children who accompanied the regiment any better off, and there was no time for delay, as the enemy were pressing closely on the rear. "The retreat was continued by forced marches amidst winter rain* and appalling difficulties." On the 29th the divisions of Generals Hope and Fraser reached Astorga, where they were joined by Baird's division, which had remained at Valencia de San Juan. Here large stores were collected, which were ordered to be destroyed. On the 31st, to reduce the strain on the commissariat, the flank brigades were ordered to separate from Moore's army and march on Vigo.

On the 1st of January, 1809, Napoleon entered Astorga at the head of 70,000 infantry, 10,000 cavalry and 200 guns; the British troops numbering at the time about 19,000 men.

From Astorga Napoleon was recalled to France by the imminence of war with Austria, leaving Soult to continue the pursuit with a force which, including the divisions of Laborde, Heudelet, and Loisson, amounted to 60,000 men and 91 guns. (Sir W. Napier.)

At this time the reserve division (Paget) and the cavalry were at Cambarros, six miles from Astorga, General Baird's division was at Bembibre, and General Fraser's at Villa Franca. That night the

* Captain Patterson says, "At intervals rain poured down with such tremendous force, that our open and straggling columns were compelled to halt, and close up in a solid body; in order that only the exterior of the mass, might be exposed to the pelting fury of the storm." (Page 85.)

cavalry and reserve marched to Bembibre, and on their arrival Baird's division marched to Calcavellos in front of Villa Franca, but, owing to the immense number of wine vaults at Bembibre, hundreds of drunken soldiers remained behind. The next morning the reserve marched to Calcavellos, closely followed by the French cavalry, and Baird marched to Herrerias, about 18 miles off. On the morning of the 4th the reserve division reached Herrerias, Baird's division being then at Nogales, Hope's and Fraser's near Lugo, where the cavalry had also been sent. On the way the 50th Regiment had a welcome opportunity of recruiting their boots and clothing from a clothing store near Villa Franca.

The discipline of the army had suffered in this retreat. The soldiers, barefooted and harassed, were dropping to the rear by hundreds, while broken carts, dead animals, and the piteous spectacle of women and children falling exhausted in the snow completed the picture.*

On the 5th of January the reserve, by a forced march, gained Nogales. Towards evening they approached Constantino, so closely pressed by the enemy that the river there was passed with the utmost difficulty; and during the night they reached Lugo, whither Baird had preceded them, and where

* Sir W. Napier gives the loss of the British army, previous to arrival at Lugo, as 1,397, and the loss from departure from Lugo to the embarkation at Corunna as 2,636.

the four divisions were now assembled. Here Sir John Moore determined to halt, for the army suffered from famine and want of rest, being compelled to retreat by forced marches, with which the commissariat could not keep up. They were burdened with a heavy weight of ammunition; at every halt they had to bivouac in the open, from which they frequently had to clear away the snow; and numbers, unable to withstand the terrific effects of cold, famine, and fatigue, sank to the earth on the bleak and barren mountains, where they speedily perished, or fell into the hands of the enemy. (Patterson.)

The 50th Regiment formed part of the army, that offered battle to the enemy for two days at Lugo. On the night of the 8th, the retreat was continued amid a terrible storm of wind and rain mixed with sleet, during which all the columns but one lost their way and at daybreak on the 9th, the rear guard was still near Lugo.

On this night, one of the outlying piquets posted to cover the retreat of the army, was furnished by the 50th Regiment under Lieutenant McCarthy. When the retreating army had marched some distance, the straw huts caught fire; and Lieutenant McCarthy so regulated the fire as to give the appearance of camp fires, and thus to prevent the enemy suspecting the retreat. Towards morning a staff officer ordered all the piquets to retire.

I 2

After retiring some distance Lieutenant McCarthy was sent back with his piquet, which thus became the rearmost troops of the army.

During the storm General Baird, having permitted the leading division to take refuge in some houses, great disorganisation followed.

The main body of the army reached Betanzos on the evening of the 9th, having suffered severely on the march. Fortunately the enemy only came up with our cavalry late in the evening.

Moore now, concentrating his army* at Betanzos, steadily retired on Corunna,† on approaching which place it became evident that the fleet had not arrived. The troops on arrival were put into quarters in the town. The land fortifications were strengthened, and those on the sea side destroyed.

On the 12th of January, the enemy's infantry appeared at Burgo, where the bridge over the River

* A soldier of the 50th Light Company was seen at Bettanzos with a fine child about 2 years old seated on the top of his knapsack. On inquiring about its parents the man stated that the mother had dropped dead on the road and he had picked up the child determined to adopt it, the father having been killed at Vimiero. He brought the child to England where it grew into a fine young man and became a shoemaker near Bury St. Edmund's.

† Patterson says, "Proceeding along the main street by the harbour side, the 50th was halted in front of a large building near the citadel, where for a short time the regiment was quartered. While we were stationed here, the great magazine of powder, situated about three miles off, was blown into the air" with an awful explosion, the sound of which reached the distant mountains.

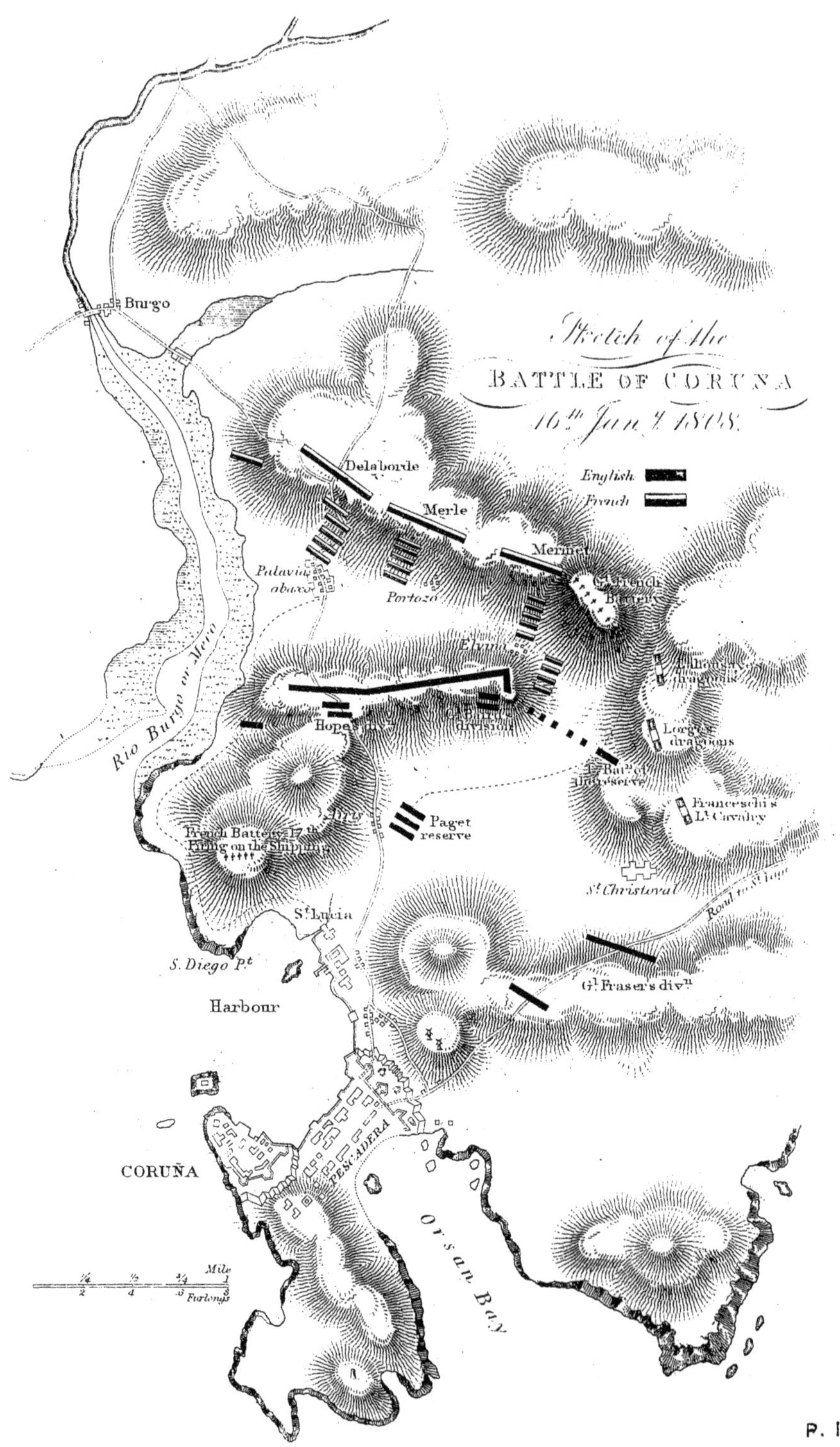
Sketch of the
BATTLE OF CORUNA
16th Jany 1808
English
French
Burgo
Delaborde
Merle
Mermet
Palavia abaxo
Portozo
Hope's div
Paget reserve
Lorges dragoons
Franceschi's Lt Cavalry
St Christoval
Rio Burgo or Mero
St Lucia
S. Diego Pt
Harbour
Gl Fraser's divn
PESCADERA
CORUÑA
Orsan Bay
Mile
1/4 1/2 3/4 1
2 4 6 8
Furlongs

Mero had been destroyed, and his cavalry lined the river's bank.

The bridge at Burgo having been repaired, three divisions of the enemy crossed over on the 14th, and that evening the transport for the embarkation of the British army hove in sight, and entered the harbour of Corunna during the night.

All the British guns except six, and the best of the horses, were embarked on the 15th ; and La Borde's division arrived to reinforce the French. During the night, Soult with great difficulty caused 11 heavy guns to be dragged to some rocks, that formed the left of his position, which was occupied by Mermet's division. Merle's division occupied the centre, and La Borde's the right. The French left, where the great battery* was posted, was 1,200 yards from the right of the British line, and midway, the little village of Elvina was held by the piquets of the 50th Regiment.

On the morning of the 16th of January, both armies faced each other on the opposing ridges; so near, that the unassisted eye could trace the slightest movement across, the intervening valley. This narrow valley was dotted with villages set amid vineyards; three of the villages were held by British piquets, and Elvina almost covered the extreme right of our position, and lay between our right, and the battery

* This was the battery that Sir C. Napier and the 50th so gallantly charged on the 16th.

of heavy guns (8 and 12 pounders) on the French left, only 600 yards off.

On the arrival of La Borde's division, the French army amounted to 20,000 men, with heavy guns, while the British forces numbered less than 15,000, with only 6 British, and 3 Spanish 6-pounder guns, worked by British gunners. About 2 o'clock, Soult opened a heavy fire from the battery on his left, which dominated the light British guns, sweeping our position to its centre, and advanced under cover of it in three columns across the valley. The first of these columns drove out the 50th piquets from the village of Elvina, and then, dividing into two parts, one part endeavoured to turn the British right, while the remainder of this column attacked in front.

At this moment, the end of the rocky eminence, that formed the British right, was held by Sir W. Bentinck's brigade of General Baird's division; another brigade of this division was in column behind the right; and one battalion, detached from the reserve, prolonged the British right on the low ground. The 50th, being the junior regiment, occupied at first the centre of Bentinck's brigade, the 4th being on their right and the 42nd on their left. This brigade, owing to the conformation of the ground, was the nearest to the enemy's position, and the most exposed to their battery of heavy guns. At the beginning of the action, Sir John Moore and staff rode up at a gallop, and he checking his horse close to where Major Charles

Napier stood, examined the enemy's advance. Being apparently satisfied, he rode off to the left.

Shortly afterwards, the right of our position being threatened by the French attack, Colonel Wynch threw back the 4th Regiment, thereby leaving the 50th Regiment on the extreme right, and Paget's reserve was ordered up to support the regiment he had previously detached.

The 50th Regiment was in line on the ridge just over Elvina; they suffered heavily from the fire of the enemy's artillery, for the range was soon found, and round shot tore through the line, and ploughed the surface of the surrounding ground.

Again, Sir John Moore came back to the 50th, just as the French column had begun to ascend the foot of the hill below them. Napier asks if he may send out his grenadier company; Moore thinks they may fire on their own piquets, but being informed that the piquets had fallen back, assents to it. The full strength of the French attack on our right being now developed, once more Sir John Moore rides up to Major Napier, and while talking to him a round shot strikes full between the two men. Sir John Moore's horse swerves away, but is forced back, and Sir John asks Major Napier if he was hit, and is answered, "No, sir." He rides away again, but soon returns and orders the 42nd to advance. Major Napier, without waiting for orders, advances the 50th in line with them. Passing the 42nd, Napier leads his regiment through Elvina, which is carried at the point of the bayonet, he cheering as he leads the advance. At this moment Sir John Moore comes back to where he had left the 50th Regiment, and, taking in the situation, rides forward in the wake of the regiment, calling out, "Well done, 50th! Well done, my Majors!"*

The light infantry company, under Captain Harrison, now charged furiously across the broken ground, and, bearing away all opposition, took lodgment in the rocks above.

The battalion companies, passing rapidly through the streets of the village, upset the cooking materials of the enemy, and, forcing every barrier, pressed on to the higher ground. "Forward! Forward to the hill!" was the cry, and clambering up the steep and craggy ascent, emboldened by the example of their officers, the men were mowed down by continuous

* Sir John Moore was engaged to be married to the sister of Major Stanhope, 50th Regiment, killed in this action. The epaulettes said to have been worn by him at the time, were presented to the regiment in 1893 by the Right Honourable E. Stanhope, M.P., Secretary of State for War in 1892, and are preserved by the first battalion. Major Napier was also indebted to the influence of Sir John Moore for his promotion in the 50th.

Captain Patterson, page 116, says, "Major Stanhope had worn a suit of new uniform and bright silver epaulettes, in which he was buried with his military cloak around him;" but as he also relates that his brother rode up at the last moment and took his ring and a lock of his hair, it is probable that he removed the epaulettes at the same time.

volleys, from the crest of the mountain bristling with French bayonets, which almost threatened to annihilate their ranks, the regiment being the most advanced in the battle, and surrounded on three sides by a sheet of fire. Major Napier's sword-belt was shot away, but looking to his front he saw the heavy battery now close above him, and gathering, by great personal exertion, about 30 men and 3 or 4 officers together, he resolved to storm the enemy's stronghold. This forlorn hope, leading straight upon the battery went down, between a fire which smote them, almost as much from their friends in rear, as from their enemies, and by the time the foot of the ascent was reached, Major Napier found himself almost alone before the enemy.

Had this dashing attempt only been supported, as Major Napier (believing the 42nd Regiment to be close behind) had every reason to anticipate it would have been, who can say, how great might have been the result; for even then Paget's division was threatening the flank of this great battery, attacked by the 50th in front.

But a series of unfortunate circumstances intervened. Sir John Moore having ordered up a battalion of the Guards, the 42nd retired, believing that the Guards had been sent to relieve them. Sir John Moore seeing this, again sent them to the front; but too late; and before the 50th Regiment could be supported, General Baird commanding the division was wounded; and Sir John Moore having received that fatal wound, which deprived the country of one of its greatest heroes, was no longer capable of giving orders for his support.

An order was sent instead to recall the 50th, and Napier and the few men with him were left alone in front of the enemy.

During this period Captain William Clunes, at the head of the grenadier company, with only a cane in his hand, compelled the enemy to evacuate a chapel, and was afterwards promoted into the 54th for bravery, and the right centre of the regiment, forcing their way through the enclosures and lanes beyond Elvina, was most severely handled, being exposed to a raking fire from the great battery, and many officers and men were killed.

Here the Honourable Major Stanhope, the second in command, met his death; and Ensigns Moore and Stewart, the officers who carried the colours, were shot down.

The latter officer was killed instantaneously, the regimental colour falling across his body, and Sergeant McKie had no sooner disengaged it, to hand it to another officer, than he also was mortally wounded!

All the ammunition being expended (70 rounds per man having been fired) and it being impossible for the regiment to maintain their position against such fearful odds, orders were given to retire; and on being relieved by the Guards the troops of the first brigade fell back, the shattered remnant of the 50th resuming their place upon the hill, from which it had at the outset advanced.

The retreat of the rest of the regiment left Major Charles Napier, with only four survivors of his party, and completely surrounded by the enemy. Gathering a few men together, he made a desperate attempt to cut his way through, but was dangerously wounded and taken prisoner, and his life was only saved by the intervention of a French drummer.* He was almost the sole survivor of that gallant little band, which he had led so bravely against the great battery: I regret that we do not know their names.† Among the many heroic deeds of the British army, I know none more deserving of being handed down to posterity, than this desperate effort of these brave men.

The heavy losses of the 50th Regiment were not suffered in vain, for no further attempt was made on Elvina by the enemy, and at the close of the action, the British everywhere maintained their position, while the French were falling back in confusion.

Preparations were now made for the embarkation

* The French drummer, Guibert, was decorated by Marshal Soult for this service.

† Private Henessey, an Irish soldier of the 50th, was taken prisoner at the same time. See Appendix for an interesting account of Major Napier's capture, taken from the "Life of Sir Charles Napier," by Colonel Sir W. Butler, to which book I am indebted for much information.

of the army, and about 8 o'clock they began to move off in perfect silence. The fires were, however, kept up, and the usual outposts maintained; those of the 50th Regiment being under the command of Captain Clunes.

A general interment of the dead took place about midnight, and Major Stanhope was buried by his brother officers and comrades, the remains being lowered into the grave with their sashes.

The 50th piquets under Captain Clunes were withdrawn an hour before the clear light of day, and marched to the shore; but the tide was so low that the boats could not get near, and the men were compelled to wade through the water above their middles to enter them. They were then embarked on "The Mary," which was anchored for this purpose so close in shore, that it made an irresistible target for the French artillery, which had by this time taken up its position upon the heights of San Lucia, supported by some infantry battalions. The captain of "The Mary," terrified by the heavy fire on his ship, cut the cable, and, as it was blowing heavily, she drifted on the rocks. The troops and crew of "The Mary" were taken on board the brig "Thomas," but with the loss of all their kit.

The 50th Regiment embarked on board the man-of-war "Ville de Paris."

The loss of the British army in this engagement is computed by Sir William Napier at 800 men; that of the French at 3,000, though he thinks the latter figure may be exaggerated.

The 50th Regiment lost—killed, Major Stanhope and Lieut. J. N. Wilson. Wounded, Major Napier and Ensigns Moore and Stewart. Killed and wounded, 5 sergeants, 4 corporals, 2 drummers, 169 privates.

The citadel was maintained by the troops under Beresford till the 18th January, when all being embarked, the fleet sailed, and arrived in England on the 23rd. The 50th Regiment disembarked at Haslar, and marched to Gosport; and from thence to Brabone Lees in Kent.

WALCHEREN.

The battle of Wagram, July 5th, 1809, having left Austria completely in the power of France, in order to create a diversion in her favour, a British expedition was prepared against Antwerp, and Flushing, which the Emperor Napoleon was endeavouring to convert into great naval depôts.

This expedition left the Downs in a gale of wind on the 28th July. During the night two of the gunboats were dismasted, and one foundered.

The fleet was under the command of Sir Richard Strachan, and the army under that of the Earl of Chatham.

The first troops landed on the island of Walcheren, without opposition, at 6 p.m. on Sunday, July 30th. The debarkation of troops was continued on the succeeding days, and the fort of Veer, having been invested, capitulated on the 1st of August.

The 50th Regiment, under Lieutenant-Colonel Charles Stewart, landed with the left wing, under the command of Lieutenant-General Fraser on the 2nd of August. They were brigaded with the 11th and 79th, and formed part of Major-General Keith's brigade.*

Lieutenant-General Fraser's division marched to Ramahim.

A fresh distribution of troops was made on the 21st of August, and the 50th Regiment was placed in the Marquis of Huntley's division, and brigaded with the 6th and 91st Regiments under Major-General Dyott.

* A battalion of detachments, nearly 1,000 strong, was raised in the Isle of Wight from the depôts of regiments on foreign service.

It was under the command of the Honourable Basil Cochrane, 36th Regiment, and included the following officers of the 50th Regiment:

Captain Henry Montgomery.
,, Edward Adkin.
Lieutenant William Turner.
,, John Patterson.
,, Richard Jones.
,, James Thomas.

It also included several non-commissioned officers of the 50th.

They were in General Fraser's division, and disembarked near the village of Veer on the 1st of August, for service in the Walcheren Expedition. Re-embarking on the 7th September, they proceeded to Portsmouth, where they landed on the 10th of that month, and from thence rejoined their respective depôts. (Captain Patterson.)

They were quartered at Oudeland and Overland in South Beveland.

That fearful sickness (afterwards known as "Walcheren fever") which devastated our army had now set in. Patterson says: "The poisonous exhalations and marsh miasmata from the loathsome waters of the canal, combined with the fervid and contaminated air, generated and extended the deadly endemic.

"Men and officers were attacked in the most sudden and violent manner while on parade, and were led away under the fatal illness, from which they were soon released by the hand of death.

"The hospitals were filled, and the convalescents reduced to so low a state, that it was a considerable time, before they were fit for any service." *

A council of war was held on the 27th August, which unanimously decided that the siege of Antwerp, the principal object of the expedition, was impracticable; and that, under the circumstances, the continuation of minor objects was not advisable.

On the 28th, the number of sick being nearly 4,000, arrangements were commenced for the embarkation of the captured guns, and the abandonment of the expedition.

Major-General Dyott's and Major-General Picton's brigades were told off on the 29th of that month to

* The number of sick on the 20th of August, 1809, was 1,564, within 3 or 4 days this number was increased to nearly 3,000, on the 27th it had increased to 3,467 and was still increasing.

hold the island of Walcheren, and for this service batteries were erected, and a floating bridge established, to enable these troops to cross over from South Beveland.

Accordingly Major-General Dyott's brigade crossed over on the 31st of August, and the 50th Regiment was quartered in the town of Middleburg.

The Queen's Own, having been especially selected to give the final outposts on the last ten days, during which the fleet was detained by contrary winds, it was the last regiment to embark, and handed over the island to the Dutch authorities in the month of December. The 50th remained in England until September, 1810, when they embarked for Portugal.

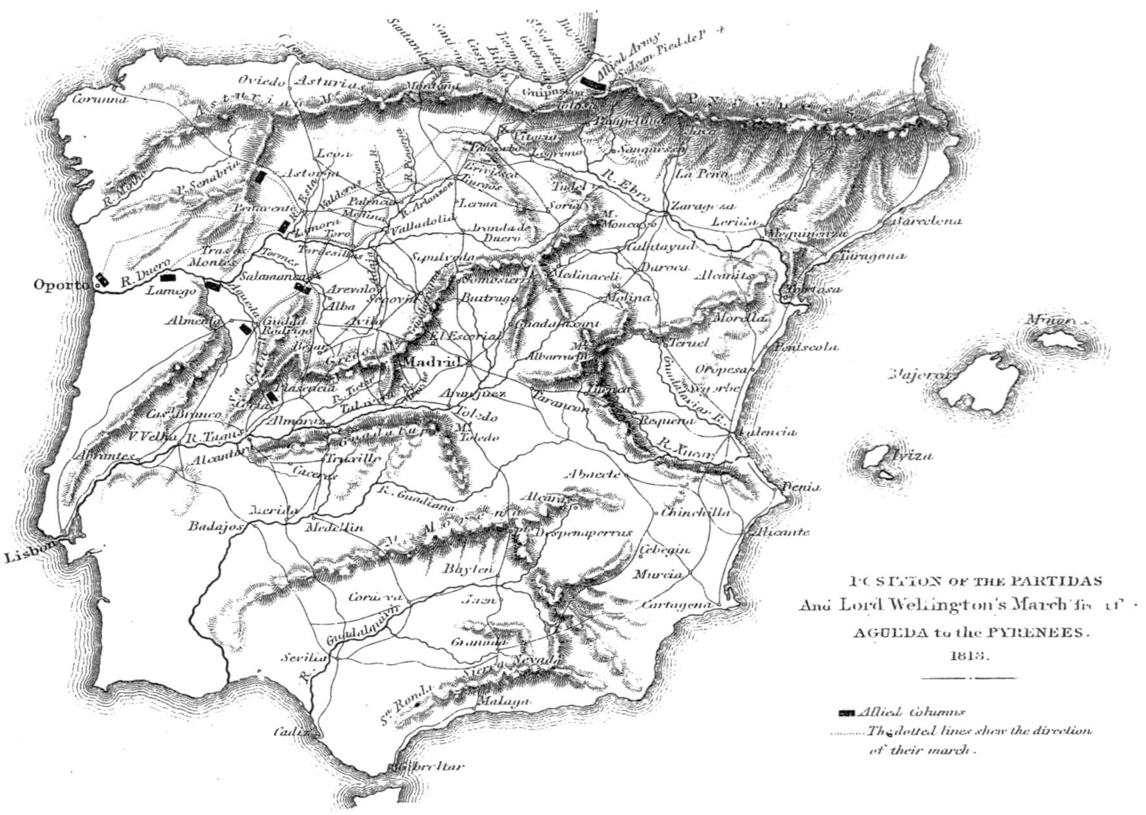
Corunna
Oviedo
Asturias
Santander
Bilboa
Guetaria
St. Sebastian
Bayonne
Allied Army
St. Jean Pied de Port
Guipuscoa
Tolosa
Pampeluna
Jaca
Vitoria
Logrono
Sanguessa
La Pena
R. Ebro
Zaragoza
Lerida
Mequinenza
Barcelona
Tarragona
Tortosa
Leon
Astorga
P. Senabria
R. Minho
Benavente
Valderas
Palencia
Medina
R. Arlanzon
Burgos
Briviesca
Lerma
Tudela
Soria
Moncayo
Zamora
Toro
Valladolid
Aranda de Duero
Calatayud
Tras os Montes
Tordesillas
Tormes
Sepulveda
Daroca
Alcanitz
Oporto
R. Duero
Lamego
Salamanca
Arevalo
Somosierra
Medinaceli
Molina
Morella
Alba
Segovia
Buitrago
Almeida
Ciudad Rodrigo
Avila
El Escorial
Teruel
Peniscola
Madrid
Albarracin
Oropesa
Majorca
Minorca
Placencia
Aranjuez
Tarancon
Segorbe
Cas. Branco
Almaraz
Toledo
Requena
Valencia
V. Velha
R. Tagus
Abrantes
Alcantara
Truxillo
Caceres
R. Xucar
Iviza
Albacete
Denia
R. Guadiana
Alcaraz
Chinchilla
Merida
Badajos
Medellin
Despenaperros
Alicante
Lisbon
Baylen
Cehegin
Murcia
Cordova
Jaen
Cartagena
Guadalquivir
Granada
Sevilia
Nevada
Malaga
Cadiz
Gibraltar
POSITION OF THE PARTIDAS
And Lord Wellington's March from the
AGUEDA to the PYRENEES.
1813.
Allied Columns
The dotted lines shew the direction of their march.

CHAPTER VI.

PENINSULA.

The First Battalion of the 50th Regiment, under the command of Lieutenant-Colonel Charles Stewart, embarked for Portugal in September, 1810. They entered the Tagus on the 25th, and disembarked at Lisbon on the 26th of that month.

They were brigaded with the 71st and 92nd Regiments, under Major-General Sir William Erskine, and formed part of the 1st Division, under Lieutenant-General Sir Brent Spencer, K.C.B.

On the 27th September, Major Charles Napier 50th Regiment, who was serving on Lord Wellington's staff, was severely wounded at Busaco.*

* Major Charles Napier, 1st Battalion 50th Regiment, who was severely wounded and taken prisoner when in command of the regiment at Corunna, was released by Marshal Ney on the 20th March, 1809, conditionally on his not serving until exchanged. In January, 1810, he rejoined the 50th. In May, 1810, he joined the Light Division in the Peninsula as a volunteer under Crawford, and was employed on Lord Wellington's staff at Busaco. He was urged to dismount, as the only man in a red coat, to which

Sir W. Erskine's brigade effected a junction, with Lord Wellington's army at Sobral, on the 10th October. This village was occupied by the 71st as an outpost, with the 50th and 92nd in support.

On the 12th October, this position was violently attacked by the enemy, the piquets driven in, and the 71st retired fighting on their supports. On the 14th, however, the French were driven back, and the original positions resumed; 200 or 300 men were wounded in these affairs.

Shortly afterwards, the brigade withdrew to Zibrera, which formed part of the lines of Torres Vedras, which position they maintained during the occupation of these celebrated lines. At this place the 50th supplied an outlying piquet in a village, at the foot of a steep precipice, beyond which, on a gentle acclivity, rose a hill, on which the piquets of the enemy were posted.

During the night of the 14th of November, the enemy retired from their position between Sobral and the Tagus; and they were followed on the 15th, by the

he replied: "No! this is the uniform of my regiment, and in it I will show or fall this day." He had hardly spoken when a bullet entered on the right of his nose, and lodged in the left jaw near the ear. He was moved to the Convent of Busaco, but in spite of his severe wound he was at Cintra next day, and from there rode into Lisbon. The 50th Regiment passed him on their way to the front, and cheered him as they passed. He was promoted to the lieutenant-colonelcy of the 102nd Regiment in June, 1811. (Extract from Life of Sir Charles Napier, by Sir W. F. Butler.) Sir C. Napier always continued to take the greatest interest in the 50th. (See Appendix, page 306, line 28.)

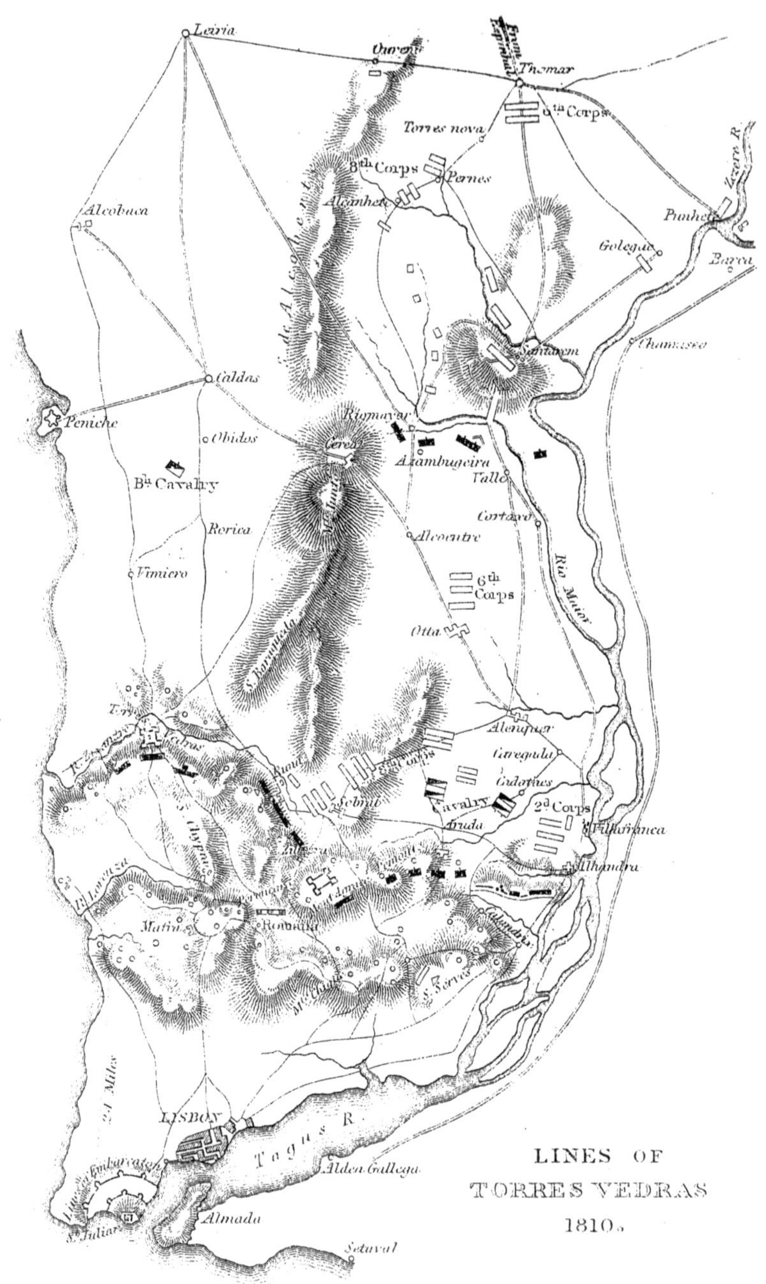

LINES OF
TORRES VEDRAS
1810.

P. 129

British cavalry and the advanced guards, as well as by Sir Brent Spencer's and the 5th Division. The enemy having, however, taken up a strong position at Santarem, the 50th and 71st Regiments were moved back to Zibrera, and the 92nd to Crozendeira, from which position the whole brigade moved at the end of the month to Alquintrinka. From there the 50th Regiment moved to Peyvallo early in March, 1811; the remainder of the Brigade being at Torre Novas.

The brigade was now commanded by Major-General Howard, Sir W. Erskine having been appointed to command a division.

The enemy retired from Santarem during the night of the 5th of March. Wellington was, however, unable to collect sufficient troops to follow till the 11th of that month.

The 1st, 4th, and 6th Divisions marched to Golegas about March 6th.

At the combat of Redinha, on the 13th of March, the 1st, 5th, and 6th Divisions were in reserve.

At Cazal Nova on the 13th, where Marshal Ney had taken up a strong position, which General Erskine had prematurely assailed with the 52nd Regiment, the 1st, 5th, and 6th Divisions, with the heavy cavalry and guns, came up in the centre in support of the Light Division, and determined the retreat of the French.

The 15th of March found Ney in a strong position on the River Ceira, near the village of Foz d'Aronce, from which they were dislodged by the 3rd Division, who were supported by the 1st; here the French lost 500 men.

The enemy having now retired to Salamanca, Wellington invested Almeida, a fortress on the Coa.

Towards the end of April, Massena, having been reinforced by a portion of Drouet's troops, made an effort to relieve Almeida.

On the 25th of that month he reached Cuidad Rodrigo; but the river Azava, a tributary of the Agueda, being in flood, delayed his advance.

Lord Wellington arrived on the 28th, and concentrated his troops behind the Duas Casas river; the Light Division being in advance along the Azava river.

On the 2nd of May, the waters having subsided, the enemy crossed the Agueda.

On the afternoon of the 3rd, they formed on the right of the Duas Casas river, and attacked the village of Fuentes d'Honor with a large force. This village, which was built on the slope of a ravine, was held by five battalions picked from the 1st and 3rd Divisions. These regiments were driven from the lower parts of the village, and with difficulty held the upper parts; and the position was becoming critical, when a gallant charge of the 24th, 71st, and 79th Regiments drove the French from the village. These three regiments were then left to hold the

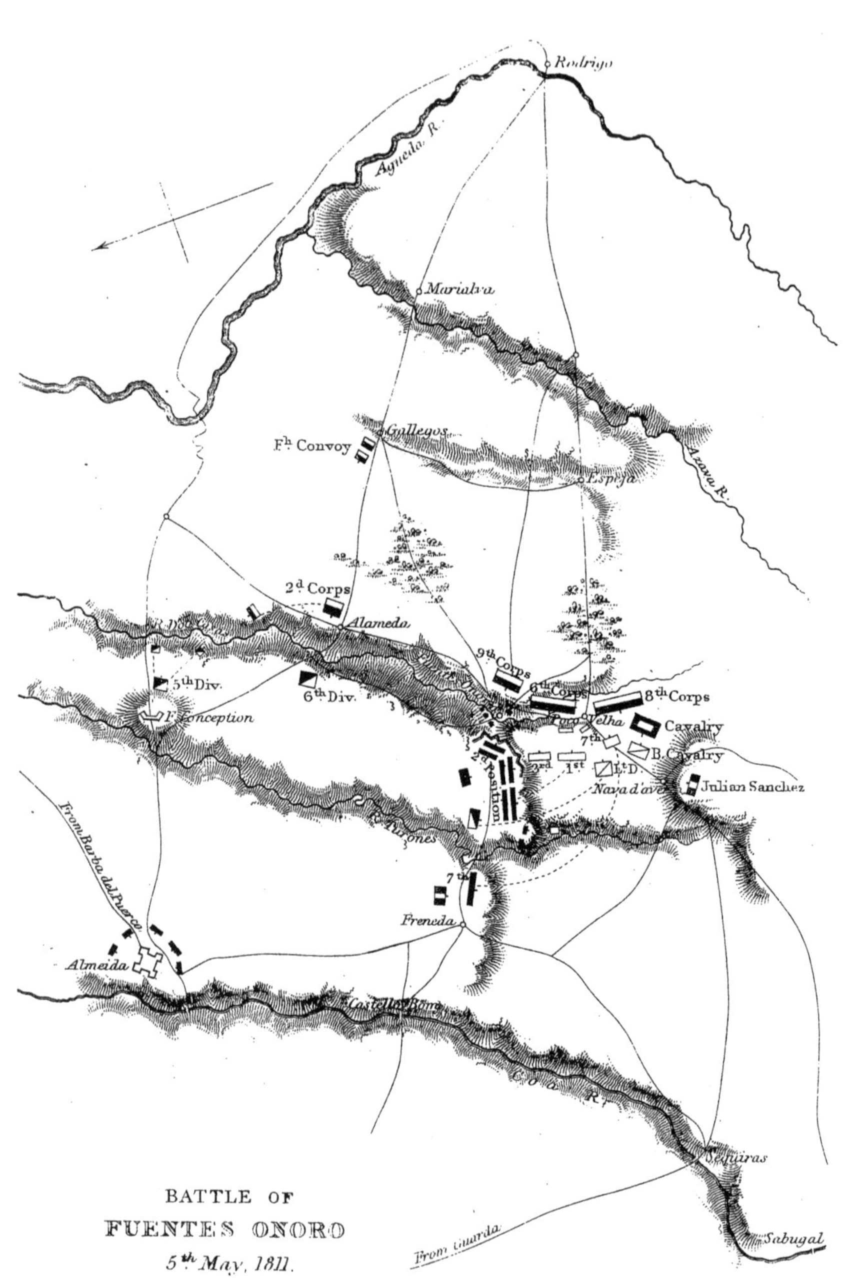

BATTLE OF
FUENTES ONORO
5th. May, 1811.

position, the other regiments being withdrawn. Lieutenant Rudkins and Ensign Grant, with 3 privates of the 50th Regiment, were wounded in this affair.*

On the 5th of May, the 8th, 9th, and 26th Corps of the enemy, with a large force of cavalry, moved to our right of the village, and the British 1st and 3rd Divisions made a corresponding movement along the ridge to their right; the bridges over the Duas Casas on the left of our position, being held by the 5th and 6th Divisions. The 7th Division was in advance near Poca Velha; but their right flank being menaced they had to retire. At this juncture the Light Division came up, and the British cavalry came into action; but the Portuguese having abandoned the hill of Neva d'Aver, our right was partially turned, the position at the time being: the 3rd Division on the left nearest the ridge, between the Duas Casas and Turones rivers, the Light Division on the right, the 7th Division slightly in advance of them, and the 1st Division in the centre. It now became necessary to throw back our right. The 7th Division was therefore ordered to cross the Turones and occupy Frenada; the Light Division covered by cavalry moved to the rear of the ridge, where they and the cavalry were in reserve; while the 1st Division, in two lines, was formed behind the ridge on the right.

* It is evident from the killed and wounded that the 50th formed one of these five battalions.

The 3rd Division was in two lines on the left, and Colonel Ashworth's brigade in two lines occupied the centre.

The enemy did not venture to assail this new position—except by a cannonade and cavalry charges* on our advanced posts—their main efforts being directed against the village of Fuentes d'Honor, of which, however, they never succeeded in gaining more than temporary possession. The 6th of May was occupied in strengthening the British position by entrenchments; but the enemy did not again attack, and retired across the Agueda on the 10th. At midnight on that date the fortress of Almeida was evacuated.

The loss of the 50th Regiment in the above action was: 3 privates killed, 2 sergeants and 19 privates wounded, and Lieutenant Ryan and 4 privates missing.

On the 22nd of May, an order was received by the 2nd Battalion, to send a detachment to reinforce the 1st Battalion.

This detachment landed at Lisbon on the 25th of June, and marched from that place on July 2nd, and on the 14th they joined the 50th Regiment, which with the rest of the brigade had been trans-

* Lord Wellington says in his despatches that one of their cavalry charges was driven back by the piquets of the 1st Division, who were afterwards themselves surprised by the enemy. As the 50th had an officer and four men missing in this affair, they probably supplied one of those piquets.

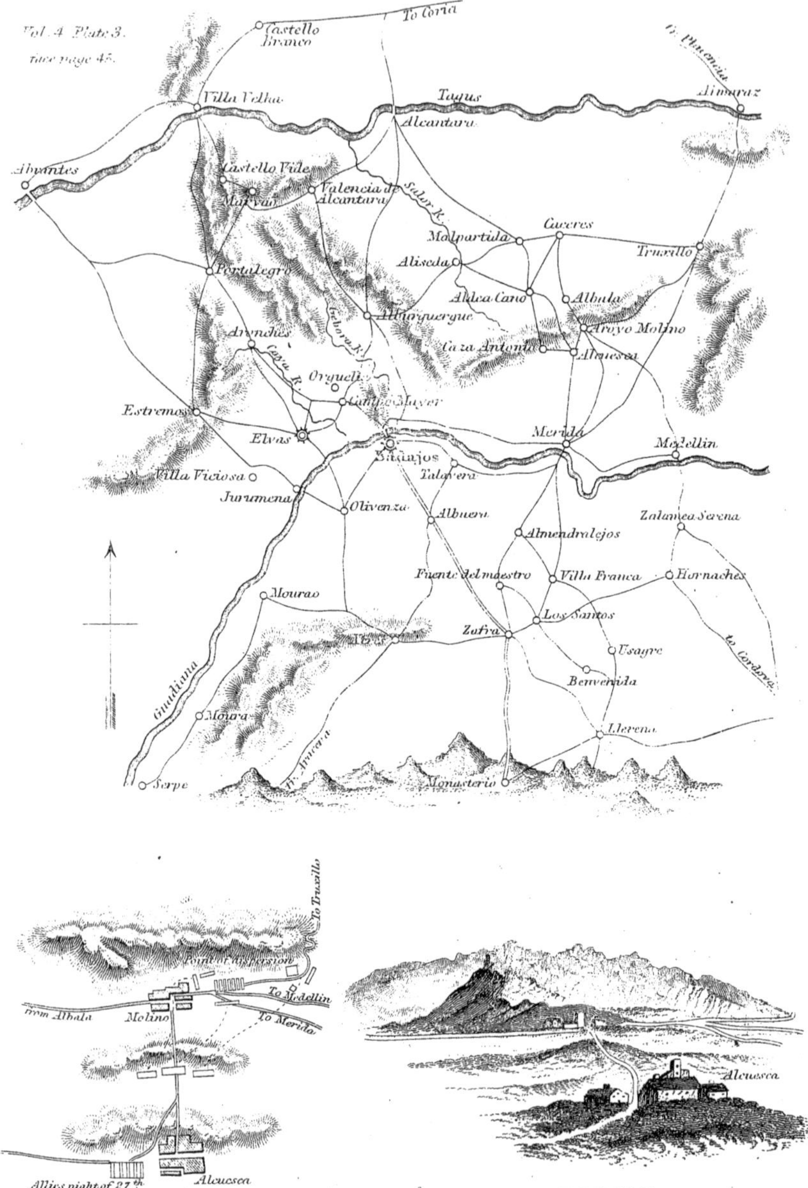

GENL HILL'S OPERATIONS.

1811.

ferred early in June, and now formed the 1st Brigade of General Hill's or the 2nd Division, and was encamped on the heights of Torre di Moro.

The troops were encamped in temporary huts, made from the branches of the oaks, which grow luxuriantly about.

On the 21st of October, General Hill received orders from Lord Wellington to march with his division towards Merida and Caceres, in order to surprise and interrupt a French corps under General Girard, and to re-open communications with the Spanish troops of La Pena and of Castanos.

The 1st Brigade of the 2nd Division were far on the road by daylight on the 22nd. They marched by the road to Albuquerque, and passing through the village of Malpartida, bivouacked in a field near the village of San Antonio on the 27th.

It was cold and wintry weather, and the rain came down in torrents all through the march, and continued with little intermission during the night; but no noise was permitted either on the march, or in the bivouac, and though there was no shelter no fires were allowed.

Before daylight on the 28th, the brigade was drawn up in the neighbourhood of Arroyo del Molinos, which was occupied by part of General Girard's force, the division being formed in three bodies, the centre one being cavalry.

The 1st Brigade, which formed the left column,

halted on some rising ground on the road close to the village, into which one company of the 60th Rifles, and the 71st were sent at the double, closely followed by the 92nd in quarter column and the 50th in close column.*

They soon gained all the principal outlets, and though the alarm was at once given by the enemy's piquets, so rapid was the movement that the surprise was complete.

Sir W. Napier says: "The horses of the rear-guard were unbridled and tied to olive trees while the infantry were gathering to form outside the village."

"The cavalry bridled up hastily, and the infantry ran to their alarm posts; but a tempest raged, a

* In the War Office Record there is a paper by Lieutenant McCarthy, 50th Regiment, which states that the 50th were at first in front of the brigade, and marching slowly were within 100 paces of one of the enemy's piquets before they were seen.

Captain Patterson, in his "Adventures," states that "the infantry, completely surprised, started from their beds, and gained an adjacent wood, on which a heavy fire was maintained." But as there is no wood shown either in the plan of this action that accompanied General Hill's original despatch, or in Sir W. Napier's plan, I have not considered this sufficiently corroborated.

Lieutenant W. McCarthy relates that a grenadier of the 50th captured a portmanteau full of dollars, and that an officer of the regiment made a most acceptable capture of a mule laden with wine and a hot ham, to which the officers did justice!

The account of this action is principally taken from General Hill's original despatch of October 30th, 1811.

thick mist rolled down the craggy mountain, a terrific shout was heard amidst the clatter of the elements, and with the driving storm came the 71st and 92nd Regiments charging down the street."

The French rear-guard of horsemen, fighting hard, were speedily driven to the end of the village, where their infantry, formed in squares, endeavoured to cover them, but the 71st, lining the garden walls, opened a terrific fire on them, the 92nd formed line across their right flank; while one wing of the 50th Regiment occupied the town and secured the prisoners, the other wing with the three six-pounders skirted outside; the artillery firing with great effect on the enemy's squares, as soon as they got within range, while the cavalry captured their artillery, and dispersed their cavalry.

General Girard, though wounded, still kept the French infantry together, and endeavoured to retreat by the Truxillo road; but the 3rd Brigade, supported by cavalry, had meantime made a rapid flank movement to their right, and were in possession of that road; and in consequence the French troops were hemmed in between the two columns, and a mountain ridge on their left, the artillery and cavalry were close on their right, and the first brigade followed fast on their rear.

General Girard's men fell by fifties, and his situation was desperate; but he would not surrender, and giving the order for the troops to disperse, he endeavoured to scale the almost inaccessible rocks.

The 28th and 34th followed them up the rocks, till they were beaten in speed by the enemy, who had thrown away arms and pack. Then the 39th and Portuguese turned the mountain by the Truxillo road; and eventually General Girard with two other generals and six hundred men, alone succeeded in joining the army of General Drouet, out of a force of 3,000 men, accounted the finest French troops in Spain; while 1,300 prisoners including Prince d'Aremberg and General Bron, all their artillery, baggage, commissariat, and a contribution just raised, fell into our hands. The loss of the allies only amounted to 70 killed and wounded.

The 1st Brigade 2nd Division then proceeded to Merida, and afterwards to Campo Mayo, from whence on November 4th they returned to winter quarters in Portalegre. On the 4th of March they proceeded to Albuquerque, and on the 22nd of that month they arrived at Dom Benito. These movements formed part of the operations for covering the third siege of Badajoz, in which the divisions of Generals Graham and Hill worked in conjunction, the former with 3 divisions of infantry, which included 20,000 Portuguese, and 2 brigades of cavalry moving on Slerena, while General Hill moved by Merida, upon Almendralejo. (See map, p. 133).

General Hill's division was afterwards encamped

in the woods, before Talavera de la Real, on the left bank of the Guadiana, about three leagues from Badajoz.*

Early in April, after the fall of Badajoz, the 50th with the rest of the 1st Brigade of General Hill's Division left Talavera, and stopping at Almendralejo and other places without anything worthy of note, arrived at Truxillo on the 15th of May.

Lord Wellington, designing to attack Marmont, it became essential to isolate his position. With this view orders were given to General Hill to destroy the bridge and forts at Almaraz on the Tagus, as the French could only cross the Tagus between Toledo and the frontier of Portugal by the boat bridge at Almaraz, to secure which strong forts and a bridge head had been constructed.

* Though the Regiment was not present at the siege of Badajoz, they were represented by Lieutenant McCarthy, mentioned in previous notes, who was attached to the 3rd Division under Sir J. Picton as leading engineer.

He was severely wounded in the assault, sustaining a compound fracture of the thigh.

The following is taken from the life of Picton :—

"Arrived in the ditch, the leading engineer, Lieutenant McCarthy, 50th Regiment, who had volunteered his services, found that the ladders had been laid on the palings of the ditch. This brave officer, finding that these palings had not yet been removed, cried out, 'Down with the palings,' and immediately applying his own hands to effect this, with the assistance of a few others, he succeeded in forcing them down. Through this gap rushed Picton, followed by his men; but so thick was the fire on this point that death seemed inevitable."

It was a bold move, as either Foy's division of Marmont's army or Drouet's force might have cut off their retreat.

On the 16th of May, Hill's force, consisting of 6,000 men, with 12 field pieces, pontoons, battering train, and 50 country carts, reached Jaraicijo, where the force was divided into three columns. The left column was to have seized the tower of Mirabette, which commanded the pass; the centre column, which consisted of the cavalry and artillery, moved by the road; while the right, which consisted of the first brigade under General Hill in person, was to move by the narrow and difficult way of Roman Gordo. The three columns started on the night of the 16th, hoping to effect a surprise, but day broke before any column reached its destination.

General Hill now decided that the Mirabette works were too strong to be carried by his force, without unjustifiable loss; yet it was only through the pass commanded by them, that he could move his artillery. He resolved, therefore, to attack the strong forts at Almaraz with infantry alone. This determination was all the more hazardous as the nearest point of retreat was Merida, four days' march distant, and it might have been intercepted by greatly superior forces.

The right column, consisting of the 50th, 71st, and 92nd, marched all through the night of the 16th, by a narrow and precipitous pathway, little

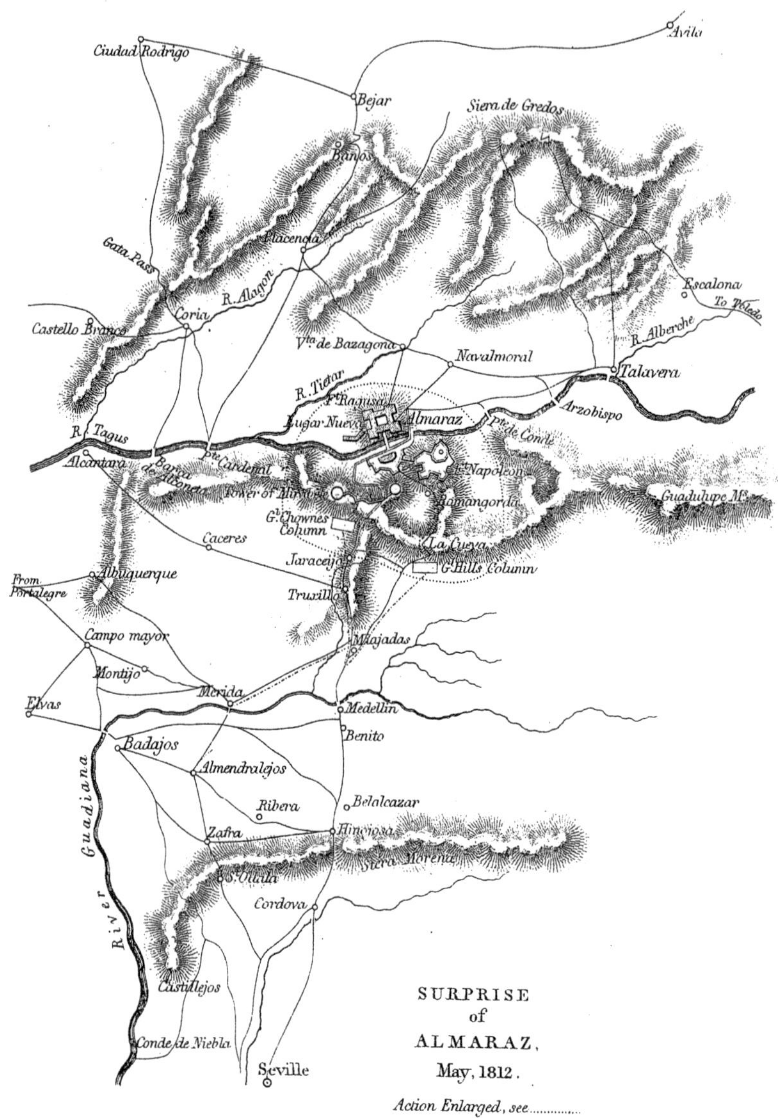

SURPRISE
of
ALMARAZ,
May, 1812.

Action Enlarged, see ....

better than a sheep walk, among broken rocks and stones, up the sides of a steep and craggy mountain.

On this mountain ridge they remained until the night of the 18th, when the advance against the forts commenced. On the south side of the Tagus, which they were approaching, the bridge over the river had a fortified head of masonry, flanked by a redoubt called Fort Napoleon, placed on a height a little in advance. Sir W. Napier speaks of this fort as imperfectly constructed, as there was a wide berme in the middle of the scarp, furnishing a landing place for escalade. It was yet strong, as it contained a second interior defence, or retrenchment, with a loop-holed stone tower, a ditch, drawbridge, and palisades. It was this fort that the 50th Regiment, assisted by five companies of the 71st Regiment, was ordered to storm.

It was intended to surprise it before daylight, if possible; but though only six miles off, owing to the difficulty of the track, the 50th Regiment, which was in advance, only reached the fort a little before daylight, while the rear was far distant; and it was doubtful whether the scaling ladders (cut in half, in order to pass the narrow turns) would reach.

The 50th Regiment, however, formed up under cover of some rising ground sloping down to the fort. Here they waited for the rear of the column to form up. It was then the morning of the 19th, and a surprise was impossible. About this time a false attack on the Mirabette works by the left column commenced; and General Lord Hill, riding up to the 50th, gave them the order to advance. The garrison of Fort Napoleon, attracted by the cannonade on the Mirabette, were crowding on the ramparts, when, in the words of Sir W. Napier, "quick and loud, a British shout broke on their ears, and gallantly the 50th, and a wing of the 71st came bounding over the low hills."

The assault was directed on three faces of the battery, for which purpose three storming parties were formed, the right wing of the regiment being led by Colonel Charles Stewart (in command), the left by Major Harrison, the third party being formed by three companies of the 71st Regiment under Major Cother, while two companies of that regiment were extended on the flanks. The whole was preceded by a forlorn hope.*

Directly the heads of the men were visible above the hill they were assailed by volleys of round shot, canister, and bullets, while thirteen pieces of canon played on them; but they advanced steadily onward under the incessant fire.

* Lieutenant W. J. Hemsworth was at first in command of the forlorn hope, consisting of 1 sergeant and 21 privates; but being wounded in the head early in the day, his place was taken by Lieutenant Patrick Plunket of the grenadier company, who survived, and was afterwards promoted into the 80th.

The order "Forward to the ditch!" was now given, and the men pressed on to the foot of the ramparts; the scaling ladders were brought up, and every effort was made to scale the walls. But the scaling ladders were too short, and for some time the effort was in vain. All this time the garrison were rolling down large rocks and stones, round shot, glass bottles, &c., on our men; and numbers were killed and wounded. Captain Candler, commanding No. 4 company,* succeeded in making a lodgment on the wall, but he was immediately killed, his body dropping inside, and his party was swept away. Lieutenant Plunket now, perceiving longer ladders in rear, with the aid of Captain North, ran back and got one, which they placed against the fort and ascended, followed by a few men. Then the grenadiers succeeded in making an entry by jumping from the ladders on to the berme, afterwards drawing the ladders up and planting them there. By a second escalade they won the ramparts; and, closely fighting, all went together into the retrenchment round

* Sir W. Napier states that Captain Candler fell leading the grenadiers of the 50th, but Captain Patterson, an officer of the regiment and an eye-witness, cannot be mistaken in saying he commanded No. 4 company, and there can be little doubt from the War Office Records, that Captain North led the grenadiers.

Captain Candler, who was promoted into the 50th from the 31st, was the only officer of the British force killed in this assault; but Captain R. F. Sandys, 50th Regiment, was so severely wounded that he afterwards died, and Lieutenant Thieb, German Artillery, was blown up by his own mine in destroying the works.

the stone tower, Captain North's party capturing two officers in the act of rallying their men. The garrison then fled in a confused mass towards the bridge head, followed by the 50th, who entered that work with them. The fugitives continued their flight over the bridge itself, still followed by the 50th, who would have passed the river had not some of the boats forming the bridge been destroyed by the guns of Fort Napoleon, which had now been turned on Fort Ragusa, on the opposite shore. Many of the enemy were drowned in attempting to cross.

Claremont, the governor of Fort Napoleon, refusing to surrender, and fighting to the last, was cut down by Sergeant Checkers of the light company. Sergeant-Major Lewis, 50th Regiment, who was conspicuous for his daring in the assault, was so desperately wounded that he died a few hours afterwards.

Meantime the 92nd, with the remainder of the 71st, who had been ordered to storm the Tête-du-pont, had been equally successful; and, the bridge of boats being speedily repaired, the 92nd dashed over it and seized Fort Ragusa on the opposite side. The total British loss in this affair was 1 captain, 1 lieutenant, 1 sergeant, and 30 men killed; 2 captains, 6 lieutenants, 4 ensigns, 10 sergeants, 1 drummer, and 117 men wounded. Of these the killed and wounded of the 50th Regiment were: killed, Captain Candler, 1 corporal, and 9 privates; wounded, Captain Sands,

Lieutenants Richardson and Hemsworth, Ensign Goddard, Sergeant-Major J. Lewis, Assistant Sergeant-Major Carlisle, 1 sergeant, 4 corporals, and 17 privates.

Two hundred prisoners fell into our hands, while the magazines turned out to be well stocked, with munitions of war as well as with provisions.

The works of Fort Napoleon and Ragusa were destroyed, and the stores, ammunitions, provisions, and boats burned. That night the 1st Brigade re-occupied the Mirabette ridge, taking with them several colours, the commandant, 16 other officers, and 250 prisoners. General Lord Hill was now about to reduce the Mirabette works, when Sir W. Erskine gave a false alarm of the French movements, which led to a hasty retreat of the whole force on Merida; and on the 20th of May the regiment advanced to Truxillo.

General Hill's Division being employed in watching General Drouet's force, did not form part of Lord Wellington's advance against Marshal Marmont, which culminated in the splendid victory of Salamanca, in which 40,000 French were defeated in 40 minutes, while Marmont himself was severely wounded, and his army was compelled to retreat across the River Douro, with a loss of 2 eagles, several standards, 12 guns, and 12,500 officers and men killed, wounded, and missing.

General Clauset, who now commanded the French

army, being still pressed, abandoned all the garrisons on the Douro and Valladolid, and retreated up the Alanza River towards the mountain. Valladolid was then occupied by the British, and found to contain large stores and 17 pieces of artillery. Wellington now crossed the Guadarama Mountains and marched on Madrid, leaving General Clinton with 12,000 men to watch Clauset, who then advanced and reoccupied Valladolid with 22,000. On the 1st of September Lord Wellington again marched against Clauset, leaving General Hill, who had been holding General Drouet in check, to hold Madrid, in co-operation with all the Spanish armies beyond the Tagus.

On the 13th of September, the 50th Regiment, who had occupied Dom Benito, forded the Guadiana about a league above Medellin, passed over the rugged hills that separate that river from the Tagus; and, passing through Toledo, they followed the course of the Tagus to Aranjuez, about seven leagues from Madrid, which they reached at the beginning of October. On the 23rd of that month they moved on the direct road to Madrid, which they skirted, the brigade being quartered on the night of the 29th at the Escurial, which proved large enough to accommodate the whole regiment, without using any of the private apartments of that splendid palace.

King Joseph now menaced Madrid with 50,000 veteran infantry, 8,000 cavalry, and 84 pieces of

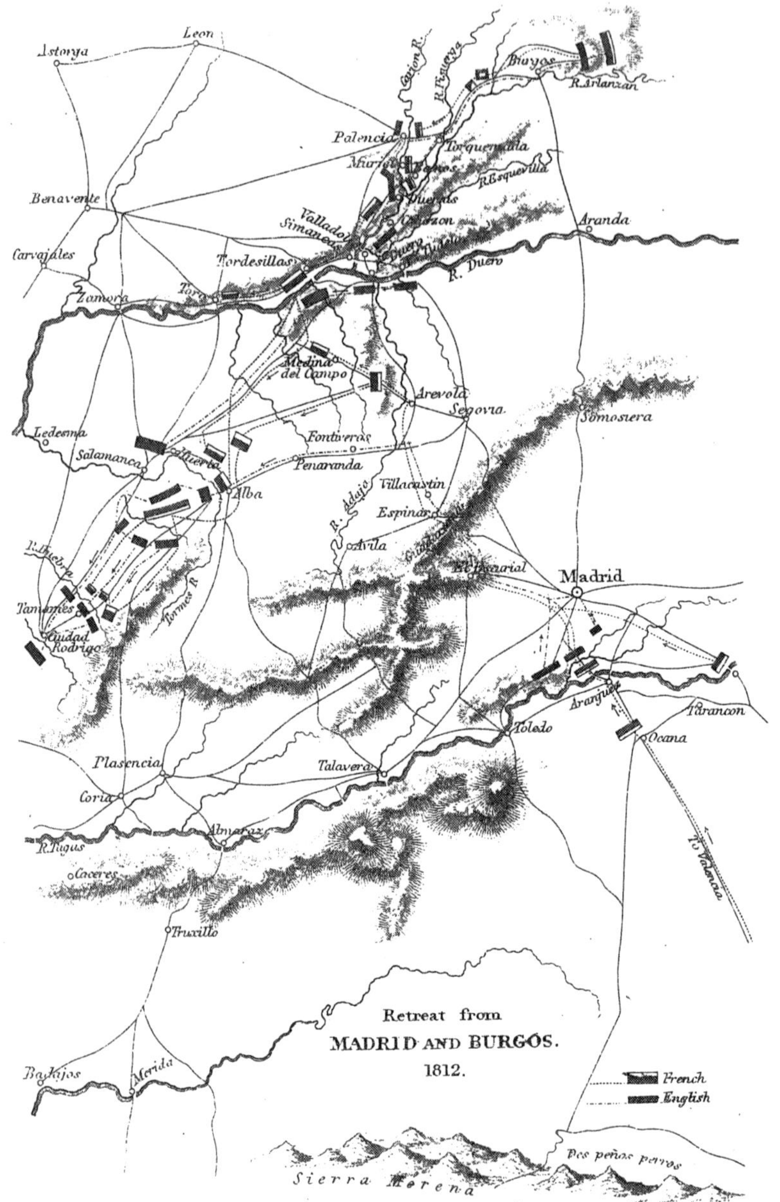
Retreat from
MADRID AND BURGOS.
1812.
French
English
Astorga
Leon
Burgos
R. Arlanzan
Palencia
Torquemada
Benavente
Aranda
Carvajales
Tordesillas
R. Duero
Zamora
Toro
Medina del Campo
Arevola
Segovia
Somosiera
Ledesma
Salamanca
Huerta
Fontiveros
Penaranda
Alba
Villacastin
Espinar
Avila
R. Adaja
Madrid
Tamames
Ciudad Rodrigo
Tormes R.
Aranjuez
Tarancon
Toledo
Ocana
Plasencia
Talavera
Coria
Almaraz
R. Tagus
Caceres
Truxillo
To Valencia
Badajos
Merida
Sierra Morena

artillery, threatening Hill's communications with Wellington through the Guadarama Pass. Meantime Clauset had fallen back before Wellington through Burgos to Briviesca, on the other side of the mountain; and Wellington, occupying Burgos, had laid siege to the castle, which successfully withstood five assaults.

Early in October, Clauset was superseded in command by General Souham, who had brought large reinforcements from France. His army now amounting to 44,000 troops, he advanced to the relief of Burgos.

On the 21st of October, Wellington, being greatly outnumbered, hastily abandoned Burgos, and retired, making stands on the Pisuerga and Douro rivers, sending word to Hill to join him in the latter position.

Early in November, therefore, Hill, having blown up the Retiro in Madrid, with all its stores, retreated by easy marches across the Guadarama Mountains; the French, uncertain of his strength, following cautiously.

The 50th Regiment, with the 1st Brigade of General Hill's Division, left the Escurial on the 1st of November to join in this movement.

Before General Hill's Division could succeed in joining Lord Wellington on the Douro, the latter found himself unable to maintain his position there; and fresh orders were sent to General Hill, who at once changed his direction, and marched on Alba de Tormes, commanding one of the passages of the River Tormes, which he reached on the 7th of November, where he formed the right of the army, Wellington occupying Christoval on the other side of the Tormes, and Calvariza Ariba in the centre. He was opposed in this position by the united armies of King Joseph and Souham,* numbering 90,000 veteran troops, 12,000 of which were cavalry, and having 120 guns.

The castle of Alba de Tormes was occupied by the 1st Brigade of the 2nd Division, under General Howard, supported on his left by the 2nd Brigade under General Hamilton.

A strong force of the enemy appeared on the heights in front of the castle, on the morning of the 10th of November, which was believed to be a reconnaissance. About 2 p.m. this force was largely increased; and 18 guns, which included six 6-inch howitzers, commenced firing on the castle, which, placed on a rocky knoll hastily entrenched, scarcely gave shelter from this tempest. Finally this post was aided by four pieces from the left bank of the river; and was defended with such gallantry that the enemy dared not attack.

General Hill says, in his despatch of November 11th, 1812: "The enemy's light troops advanced

* General Souham was afterwards superseded by General Drouet.

close to the walls, which had been hastily thrown up; but from the cool and steady conduct of the 50th Regiment under Colonel Stewart, the 71st Regiment—the Hon. Colonel Cadogan—and the 92nd Regiment—Colonel Cameron—the enemy dared not attempt the town. I feel much indebted to Major-General Howard, and also to every officer and soldier of this excellent brigade, for their steady, zealous, and soldier-like quality."

The loss of the 50th Regiment amounted to 2 privates killed, 1 sergeant and 10 privates wounded.

Soult having thus unsuccessfully cannonaded the castle of Alba de Tormes on the 10th November, crossed the Tormes on the 14th, about eight miles above that place, with the intention of turning our right, and cutting off our retreat on Rodrigo.

Leaving 300 Spaniards in the castle of Alba de Tormes, the bridge over the river being destroyed, Wellington drew off Hill's Division, which, accompanied by the cavalry and some guns, was ordered to attack the head of the French column. A race now ensued between the two columns, but, by forced marches, Hill's Division got the lead; and when Soult arrived at Tamames he found Hill in position there. Assisted by the weather, which was wet and misty, the rest of Lord Wellington's army now passed right round the flank of the French army and gained the roads leading to Rodrigo, on which they retired by three roads on the 16th. This retreat, being through vast woods in very rainy weather, was accompanied by much hardship from the incessant wet, and the difficulties of the commissariat; the troops often being glad to feed on the acorns found in the forest, and occasionally appropriating some of the pigs from the many herds fed in the woods. (Captain Patterson). A stand was made on the Huebra River, where a skirmish took place, and some of our baggage suffered from the enemy's cavalry.

On the 18th the retreat on Rodrigo was continued under great difficulties, over a marshy and flooded country, which caused great numbers of stragglers, while the wounded had to be left behind to perish.*

The 50th Regiment was now quartered in the village of Robledo, where Charles Stewart, the colonel of the regiment, died. Here they speedily recovered from the hardships of their retreat, which Sir W. Napier calculates cost the allies (including the siege and retreat from Burgos) not less than 9,000 men, with much baggage. The regiment, now commanded by Colonel J. B. Harrison, moved to the town of Coria on the 30th, where Colonel Stewart was buried.

On the 7th January the 50th left Coria and marched to the large village of Monte Hermosa, remaining there until the 8th of February, when they marched through Santivanez, Aggal, Gihon, and

* The French army did not pass the Huebra River, so the British wounded were abandoned by both sides.

Banos, to Bejar, where they arrived about the 11th of February.

The old town of Bejar, which the regiment was told off to garrison, is an advanced post some distance in front of a strong and almost impenetrable defile in the mountains of Candaleria, not far from Banos, where the rest of the brigade was quartered; and it was also opposed by a French force under General Foy at Salamanca, about twelve leagues off. It stands on the crest of a rocky eminence, and is partly surrounded by old defences containing five gates; but the walls were in a most dilapidated condition. (See map, page 138).

The 50th Regiment was at once set to work to repair (as far as could be done by loose stones and clay) the many breaches in the crumbling walls, and strong parties were told off to guard all the weak places. But the defence, which embraced the whole circuit of the scattered suburbs, was so extensive, that the utmost vigilance was necessary to guard against an attack, which was daily expected. Though the weather was bitterly cold, every morning an hour before daylight, the regiment occupied their respective alarm posts.

On the night of the 19th February, word was brought that General Foy's troops were advancing to the attack, many of the officers being at the time at a public ball. Very early on the morning of the 20th the enemy's force, estimated at 2,000 men, with some cavalry, was discovered advancing along the road from Salamanca; and as soon as it was clear daylight, a strong attack was made on a piquet of the regiment under Captain Benjamin Rowe, posted in a farmhouse on the road. With desperate obstinacy the piquet disputed every inch of ground, till, overpowered by superior numbers, they were compelled to fall back on the reserve posted near the town. The enemy now made a determined attack on the principal gate, where they were met by a well-directed fire, from the party in charge of the gate, under Lieutenant William Deighton of the Grenadiers. Many volleys were exchanged, and the French pressed on to within thirty or forty paces of the walls; but they were met there by so destructive a fire, that they were compelled to retreat; and they did not again venture to attack Bejar, during the remainder of the time the 50th were there.

The collapse of Napoleon's gigantic expedition against Russia having taken place while both armies were still in winter quarters, Lord Wellington recognised that the time had come to strike a decisive blow against France. With this view he went to Cadiz, and there obtained paramount military authority over the Spanish and Portuguese troops. He thus found himself at the head of a great army, estimated by Sir W. Napier to number about 200,000 men.

Early in May, Wellington commenced his advance by sending General Graham with a force of 40,000 men, destined to form the left wing across the Tras Os Mountains, to turn the line of the Douro by the Esla river; while he himself followed on the 22nd of the same month, with a force of 30,000 men, which included Hill's Division,* and formed the right wing—the 50th Regiment with the rest of the 1st Brigade having left Bejar and Lanos about April 17th to join in this movement. The right wing forced the passage of the Sonnes on the 26th and reached the Douro on the 28th, which they crossed eventually by a pontoon bridge. Wellington now, leaving Hill in command, crossed the Douro at Miranda by a basket slung across a deep ravine, and joined General Graham's force, which, with infinite difficulty, was forcing its way to the Esla. On the 1st of June, however, his van had crossed that river and seized Zamora on the Douro.

On the 4th of that month, the two wings having united and being reinforced by Spanish auxiliaries, the allied army (which amounted to 100 guns and 90,000 men) moved forward in irresistible strength against the scattered force of the enemy, who retired before it (Napier's "Peninsula"). The strong positions behind the Carion and Pisuerga rivers were in turn abandoned, and King Joseph took up a position

* Hill's Division consisted of the following regiments: 28th, 29th, 31st, 34th, 39th, 50th, 57th, 66th, 71st, and 92nd.

in front of Burgos, while Reille was strongly posted on the Hormoza stream, barring the road by that direction. On the 12th Wellington sent his right wing against Burgos, while with his left he outflanked Reille and compelled him to retreat. Burgos being then no longer tenable, King Joseph removed the stores, mined the castle, which destroyed 300 of the enemy in its fall, and retired to a fresh position in rear, his centre covered by the fortress of Pancorba.

Lord Wellington then by a master-stroke moved his whole army to new ground on their left near the sources of the Ebro river, masking the movement by his cavalry and by masses of Spanish auxiliaries. He thus got in rear of the French, cutting them off from the sea, and threatening their communications with France, while at the same time he shifted his base of operations to the port of Sant Andero, where a depôt and hospital station were at once established. Then for six days, Wellington pressed steadily onwards with all his divisions towards Vittoria. Great difficulties impeded the transport of his artillery through the rugged passes, but they were overcome, and it was only from a chance encounter, between Graham's force and Reille's column, in his effort to reach Ordunna, that King Joseph learnt that at least 6 divisions of the allies, were on his right and rear. A hazardous retreat of the French followed, and at great risk the King succeeded in concentrating 60,000 men in the great basin of

Battle of
VITTORIA,
with
Operations
before and after
The Action.
Allies
French
Guipuscoa
Durana
From Durango
Gamara Mayor
Betonio
Vittoria
To Logrono
To Estella
Ariaga
Armentia
Ali
From Orduna
R. Zadora
Gomecho
Hermandad
Villattes Divn.
Margarita
Arinez
P. Mena
3d. Divn.
7th. Divn.
Tres Puentes
Villodas
Nanclares
Subijana
4th. Divn.
Puebla
Miranda
Ebro R.
To Haro
Hill's corps
Espejo
St. Millan
Subijana
Morillas
Bayas R.
Bidassoa
Passage
Pampeluna
Tafalla
Tudela
St. Sebastian
Guitaria
Tolosa
Estella
Ebro R.
Deba
Azpeytia
Villa Real
Logroña
Soria
La Rioja
Mondragon
Durango
Vittoria
Domingo Calcada
Bermeo
Bilbao
Lodio
Miranda
Pancorbo
Castro
Orduña
Valmaceda
Osma
Frias
Briviesca
Santana
Poza
S. Martin
Burgos
St. Ander
Reynosa
Villadiego

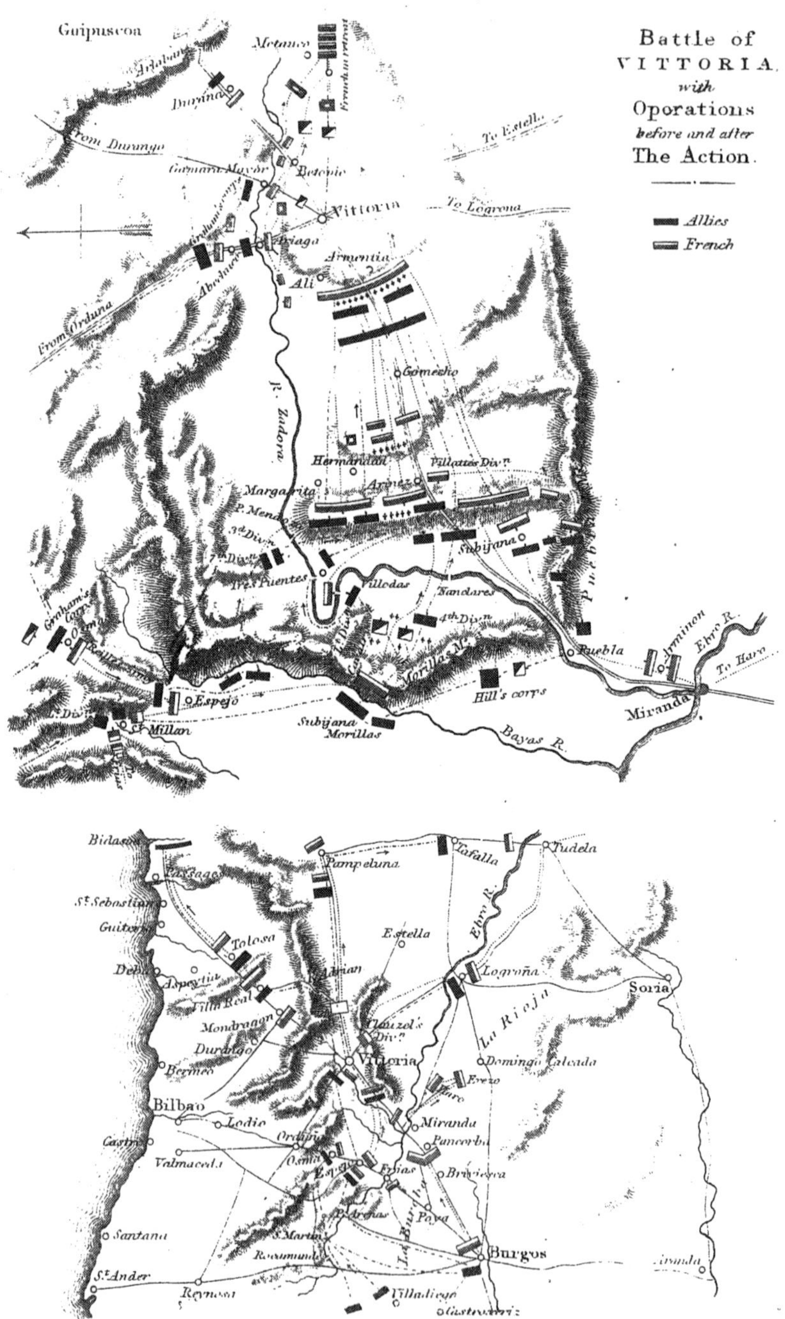

P. 153

Vittoria; which was encumbered by baggage and all the parcs and depôts withdrawn from Madrid, Valladolid, and Burgos.

The basin of Vittoria in which the French had taken up a position was eight miles broad by ten long, Vittoria being at the further end. The Zadora, narrow, and with rugged banks, after passing through that town, flows through the Puebla defile towards the Ebro, dividing the basin unequally, the larger portion being on the left bank. A traveller coming from Madrid would enter the basin by the Puebla defile, breaking through a rough mountain ridge. The town of Vittoria would then be eight miles in front; the road to Bilbao crossing the Zadora by a bridge near Ariago on his left, the Royal Road near the centre, and the road to Pampeluna on the right. (Sir W. Napier.)

The French disposition was as follows:

Reille was on the extreme right, where the Bilbao and Duranga roads crossed the Zadora by the bridges of Gamarra-Mayor and Ariago.

General Drouet occupied the centre, six or eight miles from Gamarra, lining the Zadora towards the Puebla defile, but on another front, as the stream here tends round the Margarita height nearly at a right angle towards the Puebla defile.

Gazan was on the left at the Puebla defile on an easy range of heights covered by the river. His right extended from an isolated hill in front of the village of Margarita to the Royal Road; his centre was astride of this road, and his left occupied rugged ground behind Subijana-de-Alava, facing the Puebla defile, with one brigade under Marensin on the ridge beyond the defile.

This last position was the one which Hill's division, (reinforced by Murillo's Spaniards and Sylviera's Portuguese with cavalry and guns, in all about 20,000 men) was told off to attack.

At daybreak on the 21st of June the battle of Vittoria began, the weather being rainy, with a thick vapour.

Hill, who was separated by a ridge from the British centre and left, was ordered to force the passage of the Zadora, beyond the Puebla defile. Here there were two bridges, that of Puebla beyond the defile, and Nanclares at the French end of it, facing Subijana-de-Alava. His dispositions were made to assail Marensin with his right, and, as soon as this attack was sufficiently developed, to force the pass, seize the bridge of Nanclares, and thus menace the French left. With this view, he at once commenced the passage of the river beyond the defile, and Morillo's Spaniards attacked the mountain, to the right of the great road held by Marensin. Here Morillo was wounded, and the French being continually reinforced, Hill ordered up his first brigade. They halted at the base of the heights, where the 71st, under Colonel Cadogan, was pushed forward; and though suffering heavily from the enemy's sharpshooters, and their colonel being mortally wounded, they forced their way to the summit, followed by the 50th and 92nd Regiments. Here the brigade formed column and advanced along the edge of the precipice, the 71st still leading. The French resisted stubbornly, retiring fighting from hill to hill, while every available cover was taken advantage of by their tirailleurs, and our men and officers fell fast. On one occasion the 71st took possession of a strong position, at the end of a broken promontory, where the rocks formed a natural fortress of great strength. It was commanded by a post, held by troops which were presumed to be Spanish, as they made no hostile movement. They were, however, French, and this was a ruse to get the regiment well under their fire; and as soon as this design was accomplished, such a murderous fire was opened on them, that the gallant regiment was almost cut to pieces, and the remnant was compelled to fall back on the rest of the brigade. For some time the fight was doubtful, and so strongly did the French fight, that the allies could hardly hold their ground. But eventually, the whole extent of the ridge was won, and the line of British and Spanish troops moved steadily forward till all opposition ceased.

In the meantime Hill had crossed the Zadora with his left, forced his way through the defile, and seized the village of Subijana-de-Alava. Then connecting this force with the now successful party on the mountain,* he resolutely held his position, until the success of the centre column under Wellington, caused the village of Arinez in rear of the enemy's right to be occupied. The position of the French, hard pressed in their front and on both flanks, became untenable, and they were compelled to fall back; and, their centre being driven back about the same time, the whole retired in disorder for about two miles, seeking to regain the line of retreat to Vittoria. On a height about a mile from that place, they concentrated 80 pieces of artillery, endeavoured to make a final stand, and the British pursuit was temporarily checked, until the 4th Division turned the position by seizing a hill on the French left. King Joseph now finding the Royal Road to be too much blocked by baggage, to allow the passage of his artillery by that route, indicated the road of Salvatiera as the line of retreat, and the whole force retired in that direction, hard pressed by our troops, while the British light cavalry galloped through Vittoria to endeavour to intercept them. At the borders of a marsh the enemy abandoned his artillery.

* It is not in the province of this work to detail the desperate struggle of Graham's force against Reille in the splendid advances of the centre under Wellington himself, which are so vividly described in Sir W. Napier's work, to which I am largely indebted for the incidents of this campaign.

Reille's Division, though in great danger of being cut off, eventually made good their retreat, and acted as a rear guard from Metauca, to which place the enemy were closely pursued.

The spoils on our side were immense; French carried off but two pieces of artillery. Marechal Jourdan's baton, a stand of colours, 143 brass pieces, all the parcs and depôts from Madrid, Valladolid, and Burgos, and all the carriages, ammunition, and treasure, fell into our hands.

The loss of the 50th Regiment in the above action consisted of—killed, 27 privates; wounded, Captains A. Gordon and J. Gardiner, Lieutenants Bower and Turner, Ensigns Williams and Reed, and 70 privates.

King Joseph now sent Reille's corps to the Bidasoa, and marched with Gazan and D'Erlon's troops on Pampeluna, which they reached in so disordered a condition that the governor refused to allow them inside.

The 50th accompanied General Hill's Division to Pampeluna, which was invested by the division on the 22nd; and on the 28th the King fled into France by the Roncesvalles pass.

Early in July, without abandoning the investment of Pampeluna, Hill detailed a part of his force (which included the 1st Brigade) into the valley of Bastan; and the remainder of his division was afterwards relieved in the blockade by Spanish troops.

Soult having superseded King Joseph in the chief command, united the three beaten armies in one under the title of the Army of Spain, and made preparations to advance, with over 100,000 men, to relieve the invested fortresses of Pampeluna and San Sebastian.

The principal passes through which he would have to advance were those of Roncesvalles and Col de Maya, the latter being held by the two British brigades of General Hill's Division under Brigadier-General Stewart, while Campbell's Portuguese preserved the communication with Byng's and Morillo's Brigades, guarding the Roncesvalles Pass on our right.

The Col de Maya was menaced by D'Erlon's corps from Espalette, and Urdax furnishing 18,000 bayonets.

The 1st Brigade marched into Elisonda on the 8th of July, and took up a position on an elevated ridge about a league in front of that place. The 71st and 92nd Regiments were encamped near the pass of Maya to the left of the road to Urdax and Espellette; and the 50th bivouacked among some trees about a mile to the right. There were two smaller passes to our right, Lessessa and Aratesque, the whole occupying about four miles.

A piquet of 80 men was stationed at the latter pass supported by 4 light companies, a mile in rear of the reverse slope.

At dawn on the 25th July, a glimpse had been

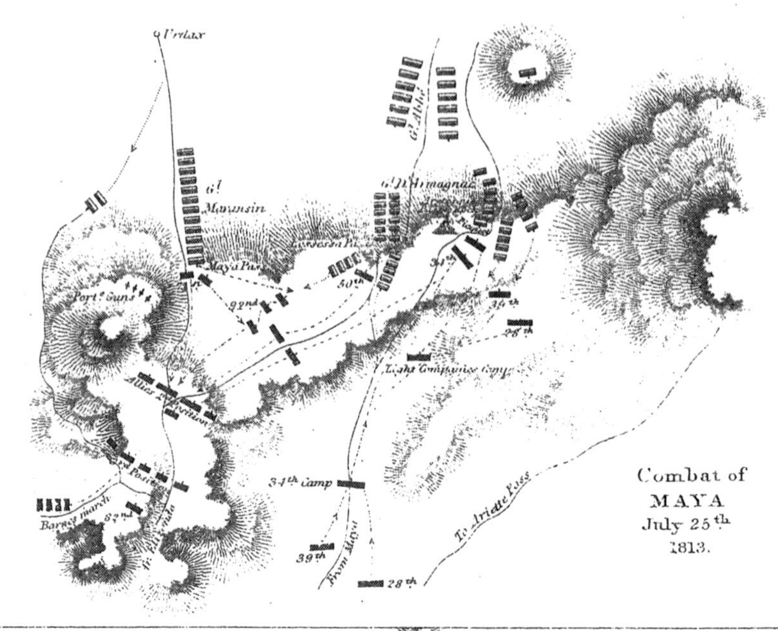

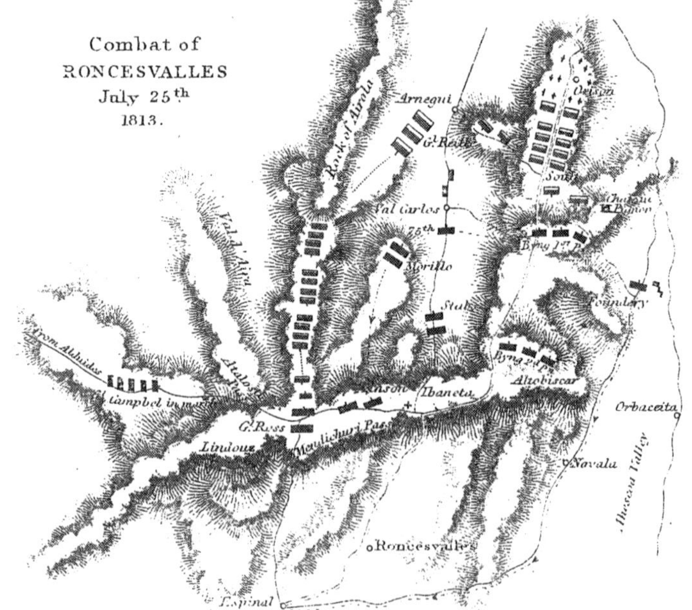

Urdax
Gl. Maransin
Maya Pass
Port. Guns
92nd
50th
34th
28th
Light Companies Camp
34th Camp
39th
From Maya
28th
To Arriette Pass
Barnes march
82nd
Combat of
MAYA
July 25th
1813.

Combat of
RONCESVALLES
July 25th
1813.
Arnegui
Rock of Airola
Gl. Reille
Orisson
Soult
Val Carlos
75th
Morillo
Val d'Aira
Foundery
From Alduides
Altobiscar
Ibaneta
G. Ross
Lindouz
Mendichuri Pass
Orbaceita
Novala
Roncesvalles
Espinal

P. 159

obtained of cavalry and infantry moving along the hills in front, and at 9 o'clock the 4 light companies were ordered to support the piquet. They had barely formed up with their left resting on the rock of Aratesque, when D'Armagnac's Division mounted the hill in front, followed by Abbe's Division, while Marensin's Division attacked the passes on our left.

D'Armagnac's Division, pushing forward in several columns, forced the piquet back, and the light companies sustained the vehement assault with infinite difficulty. The alarm guns were now fired from the Maya Pass. The 34th was the first regiment to come up, by companies, followed later and on lower ground by the 39th and 28th. Meantime the 1st Brigade formed up on the highest part of the ridge, near the Col de Maya.

The 34th, though fighting splendidly in conjunction with the piquet and light companies, were unable to hold the ground against the overwhelming numbers of the enemy, and were forced back. Colonel Fenwick, their commander, was severely wounded. The French were thus left in possession of this pass, but Cameron,* seeing that the 34th were hard pressed, sent the 50th Regiment to their assistance. They surprised a part of D'Armagnac's

* Colonel Cameron, commanding the 92nd, was in temporary command of the brigade, as Brigadier-General Stewart had not arrived from Elisonda.

Division, just as they had gained the summit of the Lessessa Pass in the centre, and, immediately charging, drove them clean out of the pass.* They then joined the 34th Regiment, and these regiments poured a destructive fire on the advancing foe. But Abbe's Division having arrived, they were again compelled to fall back, and the 50th Regiment now retired on the position taken up by the 92nd Regiment.

Here they formed in line † in front of about 5,000 of the enemy, who advanced in contiguous columns. The Regiment, undaunted by numbers, charged, and crossing bayonets with the confident foe, threw them into confusion. The loss of the regiment was so severe, that it was obliged to retire a short distance, when it again charged, crossing bayonets a second time with the imposing column, who, confident in their great numerical superiority, rallied after each charge, and kept up such a destructive fire as reduced the regiment in a short time to about half the number it brought into action. The regiment again retired

* Sir W. Napier, writing of this charge of the 50th, says: "That fierce and formidable old regimont charging the head of an advancing column, drove it clear out of the pass of Lessessa.

† The description of the subsequent splendid charges of the regiment from "here they formed in line" is taken almost verbatim from a document in the War Office.

Patterson says the 50th were supported by the 39th; but this must be an error, if Sir W. Napier is right in saying "the 39th and 28th were cut off from the others, and forced back to a second and lower ridge crossing the main road to Elizonda."

in the steadiest manner about 200 paces, and it was again preparing to charge when an order was received to retire to a rocky ridge a little in rear, where it was followed by the 92nd Regiment, who had lost two-thirds of their numbers, in this gallant struggle.

Finally, all the positions in front were lost, and the four Portuguese guns on our left taken, General Stewart taking up a second and eventually a third position in rear; but it was not until the arrival of General Barnes's Brigade of the 7th Division, that even this last position was secure.

The Regiment suffered very severely in this action, especially the grenadier company. Captain Ambrose, who commanded them, was mortally wounded; and, calling his sergeant, gave him his sword, saying, "The enemy shall never have it!" His subaltern, Lieutenant Deighton, was also killed with his cap on his sword, cheering on the men. Lieutenant-Colonel Hill was severely wounded after the second charge, and the temporary command of the regiment devolved on Major J. D. Campbell.

The total loss of the regiment was—killed, Captain Ambrose, Lieutenants H. Burchill, and Deighton, Ensign White, and 40 privates; wounded, Lieutenant-Colonel Hill, Captains Grant, North, and Rudkin, Lieutenants Nowlan, Patterson,* Jones,

* Captain Patterson, the author of "Adventures of Captain John Patterson," frequently referred to in this work, was

Plunket, Crofton, Ensigns Richards and Bateman, and 90 privates.

That night, the 25th, Hill collected his division and retired to Irueta, fifteen miles from the scene of action, leaving a Portuguese brigade in front of Elizonda. In this position he covered the roads to San Esteven and Berderes, and the pass of Vellate in his rear. This position he maintained till the 27th, when he marched during the night through the pass of Vellate to Lanz, which he reached on the 28th, and marched the same day to Lizasso, in rear of which he took up a position on a ridge covering the road to Marcalain. (See map, page 159.)

Meantime the pass of Roncesvalles had been abandoned after hard fighting, and Soult advancing in great strength with his main army, Picton retired before him. Eventually, assisted by Cole's Division, and by Spanish and Portuguese auxiliaries, he took up a strong position on a rocky ridge covering Pampeluna. On the 27th Soult made an attack on a hill on our right of the position held by Spanish troops, but was repulsed.

On the 28th Wellington brought 6,000 men into the allies' line of battle, covering Pampeluna, and 15,000 more into military communication with their left.

On the 28th Soult vehemently attacked the allied

wounded in the above affair, and his valuable descriptions of the actions of the regiment cease from this date.

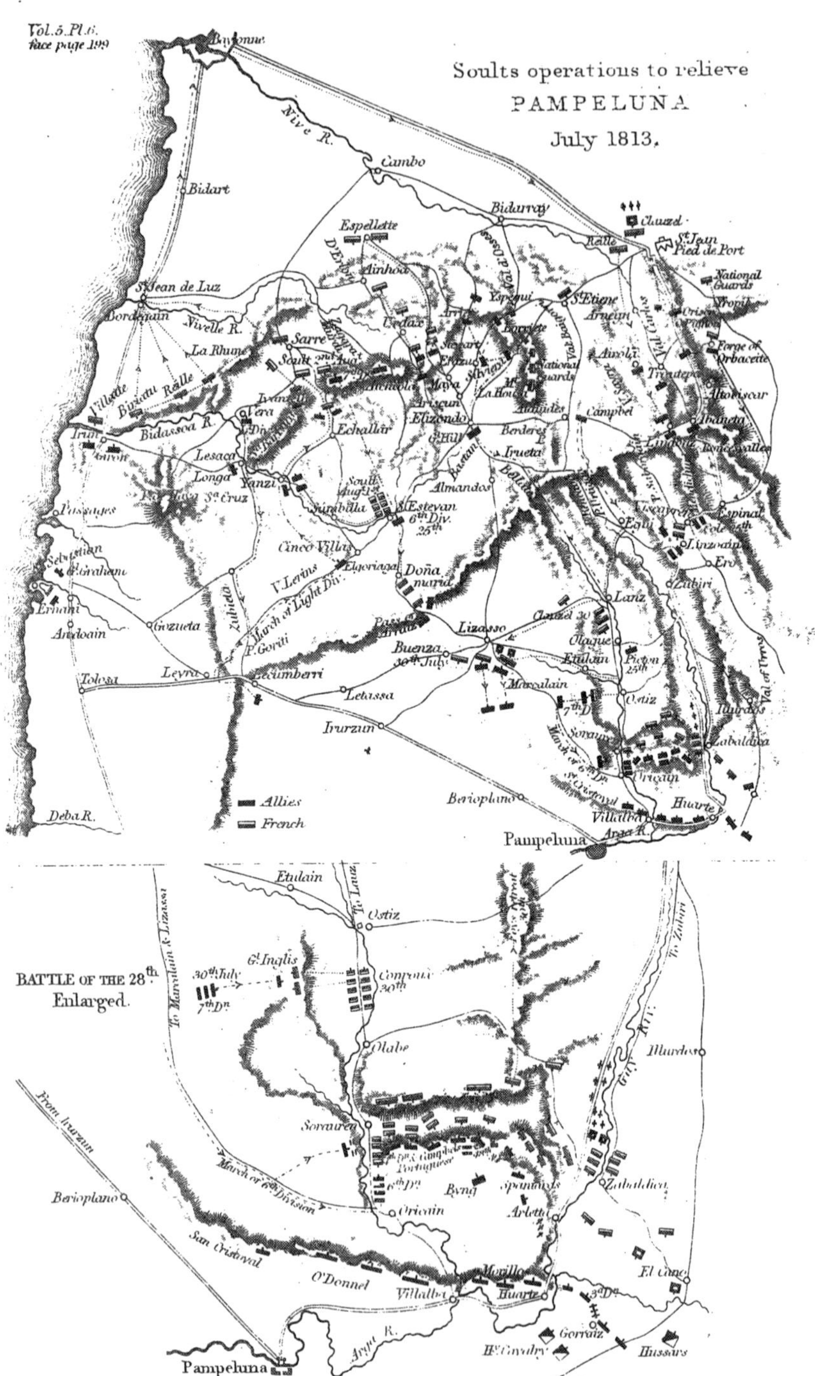
Vol.5.Pl.6.
face page 199
Soults operations to relieve
PAMPELUNA
July 1813.
Bayonne
Nive R.
Cambo
Bidart
Bidarray
Clauzel
St. Jean Pied de Port
Espellette
Ainhoa
St. Jean de Luz
Bordegain
Nivelle R.
La Rhune
Sarre
Urdax
Espegui
St. Etiene
National Guards
Forge of Orbaceite
Altobiscar
Ibañeta
Roncesvalles
Maya
Elizondo
Echallar
Bidassoa R.
Irun
Vera
Lesaca
Yanzi
St. Cruz
Almandos
St. Estevan
6th Div.
Campbel
Linzoain
Espinal
Zubiri
Erro
Passages
Cinco Villas
Doña maria
Lanz
Zubieta
Gozueta
Andoain
Lizasso
Buenza
Ostiz
Marcalain
Olaque
Etulain
Picton
Tolosa
Leyra
Lecumberri
Letassa
Irurzun
Sorauren
Oricain
Zabaldica
Huarte
Villalba
Arga R.
Berioplano
Allies
French
Deba R.
Pampeluna
BATTLE OF THE 28th. Enlarged.
Etulain
To Lanz
Ostiz
G.l Inglis
30th July
7th Dn
Olabe
Ilurdos
To Marcalain & Lizasso
From Irurzun
Sorauren
March of 6th Division
Berioplano
6th Dn
Byng
Spaniards
Oricain
Arleta
Zabaldica
San Cristoval
O'Donnel
Morillo
Villalba
Huarte
El Cano
3rd Dn
Gorraiz
Arga R.
Pampeluna
Hussars

position, meeting with a temporary success on our left, owing to the giving way of a Portuguese battalion, but, assisted by the opportune arrival of the 6th Division from Lizasso, the allies succeeded in beating off the attack, and in holding their position, although only with desperate fighting, and with heavy losses on both sides.

On the 29th both armies occupied their respective positions, and Hill sent all his baggage, artillery, and wounded to Berioplano, on the great road to Tolosa.

On the 30th Soult, being unable to supply his army so far from his depôts, commenced a retreat towards the valley of Bastan; and, to cover it, D'Erlon, who with his four divisions had remained inactive after forcing the pass of Maya, was ordered to advance by Lanz and Etalain to Lizasso, which, it will be remembered, was covered by Hill's position in rear of the Marcalain ridge. Thus the same troops, who met at the pass of Maya, were again opposed. On arrival, D'Erlon at once made preparations to attack. Hill's right was strongly posted on rugged ground; but his left, prolonged towards Buenza, was insecure. D'Armagnac's Division was ordered to make a false attack on his right. Abbe's Division, followed by Maransin, was to turn his left and gain the summit of the ridge. The cavalry commenced these attacks, and La Martiniere's Division, coming from Lanz, brought up the rear. Soult was present in person. D'Armagnac, pushing his feint too far, became seriously engaged, and was beaten by the Portuguese and 28th Regiment; while on our left the French divisions were repeatedly repulsed. Abbe, however, eventually gained the summit of the mountain, which turned our left, and, the position being no longer tenable, Hill threw back his left and retired to the height of Yguaras between Arestegui and Berasin. There, uniting with Campbell and Morillo, he again offered battle.

In this affair Hill lost 400 men, and the 50th Regiment once more distinguished themselves by a charge, in which they for a third time crossed bayonets with the enemy, and in which Lieutenants Bartley and Power were taken prisoners.

In the meantime, the capture of Sorauren, and the success of the allies under Picton on our right, driving back the French left, together with Picton's rapid advance, had isolated Soult's forces; and the position taken up by Hill behind Marcalain (which was now also supported by the 7th Division) menacing his retreat by Lizasso, he was compelled to retreat by the Donna Maria Pass. During the night of the 30th he commenced his retreat, leaving D'Erlon's Division to follow as a rearguard.

On the 31st Hill followed with the 1st and 7th Divisions; and at 10 a.m. he overtook the rearguard, part of which only succeeded in gaining a wood on the summit of the pass under fire of his guns. General Stewart at once attacked this position with the 1st Brigade of the 2nd Division, when the 50th Regiment were again engaged; but General Stewart was wounded, and the attack of the 1st Brigade failed. Later, however, the position was taken by the 2nd Brigade; and the 7th Division having also been successful on the right, the enemy retreated. A thick fog soon put an end to the pursuit.

Captain Wemyss, 50th Regiment (acting Brigade Major), was killed here; and Hill's Division lost close on 400 men.

In accordance with orders Hill now fell back to Lizasso, from whence he marched to Almandoz, leaving the 7th Division to continue the pursuit.

Soult halted at Estevan on the 1st of August, and Wellington made strenuous efforts to surround his column.

Wellington had 3 British and 1 Spanish division behind the mountains overlooking Estevan. General Alten, with the Light Division, had received orders on the 31st, and, in conjunction with Graham's Spaniards, was marching to intercept the head of the French columns, at Vera and Echallar. The 7th Division was in their rear at Donna Maria. Byng was at Maya, and Hill was marching by Almandos to join him.

On the 1st of August, before there was time to complete these dispositions, most unfortunately, three stragglers were captured, who revealed the position of the allies; and within half an hour Soult's Division was on the march, by Sumbilla and Yanzi. After nineteen hours' consecutive marching, the Light Division reached his leading column at Yanzi; and towards evening the bridge was taken and the route barred. The confusion in the narrow and rocky valley was indescribable, and the French loss immense; yet they were able to escape by Sumbilla to Echallar. Nor did Soult's rearguard under Clauset fare much better, for the skirmishers of Cole's Division and O'Donnel's Spaniards opened a heavy fire upon them from the height above Estevan, which he was unable to return. His baggage was taken, many of his troops were dispersed, and it required the personal exertions of Soult himself to prevent a total rout.

That night Soult rallied his divisions about Echallar, his right at Ivantelly communicating with Villate's Division near the Great Rhune Mountain.* The next morning his left, under Clausel, was driven back by the splendid action of General Barnes's Brigade of the 7th Division; and he fell back and took up a new position beyond the Pass of Echallar, his right resting on the Ivantelly Mountain.

The position was very strong; but, without waiting for reinforcement, 5 companies of riflemen, supported by 4 of the 43rd, fought their way up

* Villate's Division had been watching St. Sebastian from Irun, on the opposite side of the Bidasoa river.

without a check to the summit, and the night fell on the French retiring in disorder.

Wellington had now succeeded in driving back the army intended for the relief of Pampeluna and San Sebastian with great loss; and Soult, having taken up a permanent defensive position behind the Bidasoa, Wellington made a corresponding movement, in which Hill's Division held the Alduides and the Roncesvalles Pass, and both armies rested from their labours.

Lord Wellington then turned his attention to the sieges of Pampeluna, and San Sebastian.

The battering train for the latter, arrived from England, on the 19th of August, and the works were taken on the 31st, after a most sanguinary assault,* the garrison retreating to the strongly fortified position of Monte Orgullo. On the 9th of September this was surrendered.

The day San Sebastian was taken, Soult, who was kept informed of the progress of the siege from the sea face, (which in the absence of a fleet could not be properly blockaded), made an attempt to relieve it,

* Three generals — Leith, Oswald, and Robinson — were wounded in the trenches, Sir R. Fletcher (chief engineer) was killed, Colonel Burgoyne (second engineer) wounded; and the carnage at the breaches was appalling. The volunteers, though brought late into the action, lost nearly half their numbers; most of the regiments of the 5th Division suffered in the same proportion; and the whole loss since the renewal of the siege exceeded 2,500. (Sir W. Napier.)

and a fierce struggle took place, principally on the St. Marcial range covering the fords of the Bidasoa, which was repulsed at all points. Hill's Division, though not engaged, was directed to show the head of their column towards St. Jean Pied de Port, thus menacing the enemy's flank.

Early in November, Wellington determined to attack Soult's position. With this view, Hill's Division was moved to Bastan on the 6th and 7th of November, Minar being left to hold the Roncesvalles Pass.

Ninety thousand combatants, and 95 pieces of artillery, were prepared for the attack on the 10th. Of these 26,000 men and 9 guns were under Hill,* who was ordered to attack the left of the French position. Beresford and Alten's columns of attack were on his left, and Sir J. Hope's column held the French right in check in front of Bayonne.

During the night of the 9th Hill moved by the different passes of the Col de Maya, and, after a long and difficult march, he neared the enemy a little after 7 a.m. on the 10th, near the bridge of Amotz. Sending Morillo and Minar against the fortified spur, extending from Mount Atchulegue to Mount Mon-

* General Hill had under his command the 2nd and 6th Divisions, Hamilton's Portuguese, Morillo's Spaniards, 4 of Minar's battalions, Grant's Brigade of Light Cavalry, and 9 guns.

Sir J. Hope, who was second in command to Lord Wellington, had recently taken over Graham's Division.

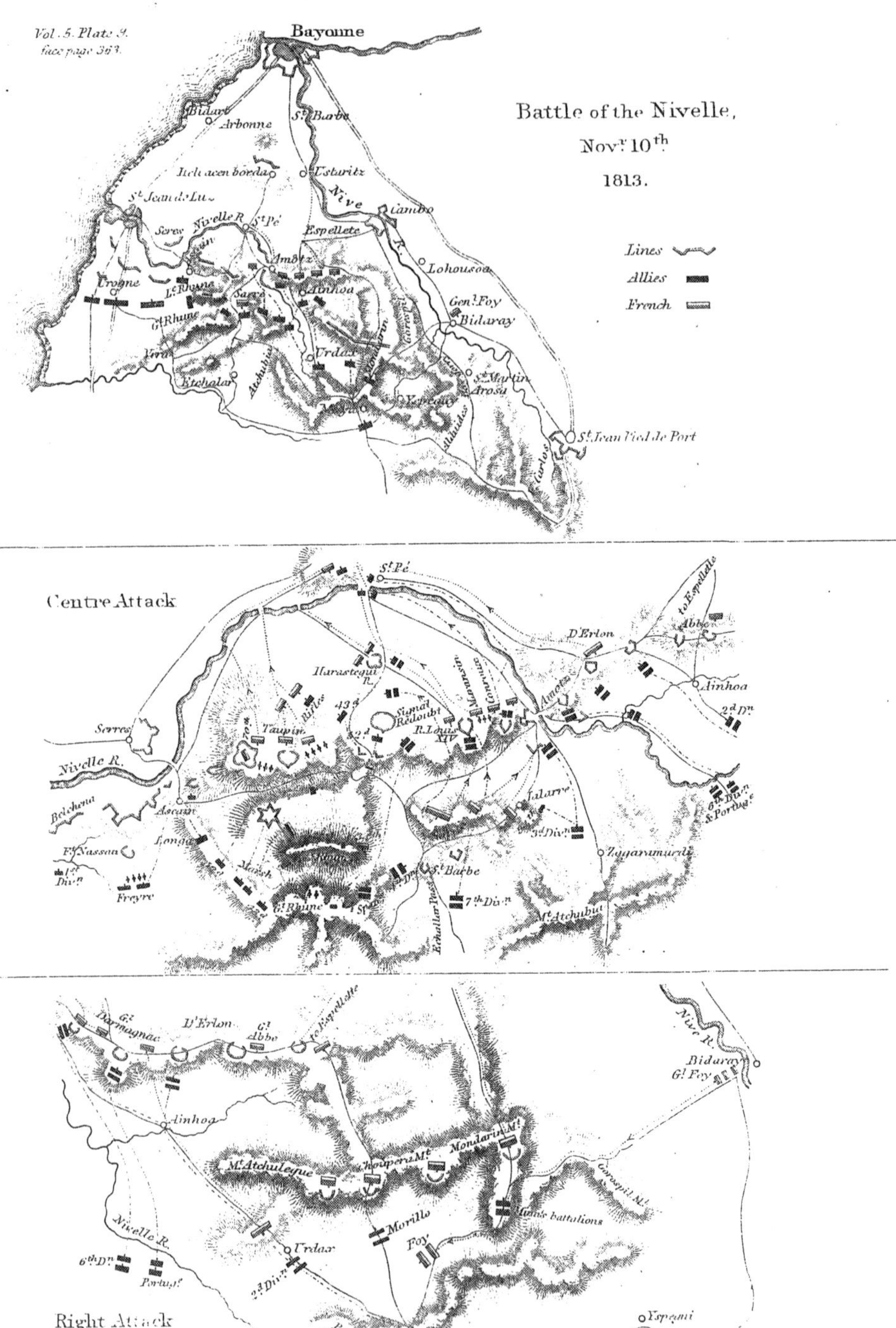
Vol. 5. Plate 9.
face page 363.
Battle of the Nivelle,
Novr. 10th
1813.
Lines
Allies
French
Bayonne
Bidart
Arbonne
St. Barbe
Ustaritz
St. Jean de Luz
Serres
Nivelle R.
St. Pé
Espellete
Nive
Cambo
Lohousoa
Amotz
Ainhoa
Sarre
Urdax
Genl. Foy
Bidaray
Gd. Rhune
Vera
Etchalar
St. Martin Arosa
St. Jean Pied de Port
Centre Attack
D'Erlon
Abbe
to Espellette
Harastegui R.
Signal Redoubt
R. Louis XIV
Taupin
Rifles
Nivelle R.
Ascain
Longa
Ft. Nassau
Freyre
Zagaramurdi
St. Barbe
7th. Divn.
3rd. Divn.
Mt. Atchubia
Right Attack
D'Erlon
Gl. Abbe
Ainhoa
Mt. Atchuleque
Choupera Mt.
Mondarin Mt.
Morillo
Foy
Urdax
6th. Dn.
Portug.
Nivelle R.
Pass of Maya
Ysperui
Nive R.
Bidaray
Gl. Foy

darin, where they held the enemy in check, he advanced with the 2nd Division and drove the French from Urdax and Ainhoa, while the 6th Division, with Hamilton's Portuguese on their right, passed the Nivelle river to Hill's left, and threatened the bridge of Amotz. The Spanish troops continued to occupy the attention of the French on the Atchulegue ridge, while an attack was made on the redoubts beyond, which were placed along the crest of a ridge thickly covered with bushes, and having a deep ravine in front. Clinton, with the 6th Division, turned this ravine on the left, and drove the enemy from the half-finished work covering the bridge, and then wheeled against the redoubt on the French right of this position, which was thereupon abandoned. Hamilton menaced the second redoubt, and Hill, with the 2nd Division, stormed the third. The French then, setting fire to their camp, retreated towards San Pè, followed by the 6th Division; and the French brigade defending Atchulegue ridge retreated towards Espelette and Cambo, followed by the Spaniards. The 50th Regiment was in support in the above affair. Meantime the columns on Hill's left had carried the positions assigned them with splendid gallantry; and he was now in communication with Beresford by the bridge of Amotz. The bridge in front of San Pè had been seized by the 3rd Division; and though the French still endeavoured to hold on to a strongly fortified ridge in the centre, they were not able to stand before the Light Division. They were compelled to retreat across the Nivelle by the fords in rear of Harastegui, and by the bridge in rear of Ascain, having the fortified camp of Serre behind it. Hither Soult had hurried, on the first alarm, with all his reserve artillery and spare troops; but he was held in check by Sir J. Hope's column.* Wellington then, leaving Cole, Alten, and Giron to watch the bridge near Serres, as soon as the 6th Division was sufficiently advanced, crossed the bridge at San Pè with two divisions, and drove the French from a strong position in which they had rallied in rear of that bridge. The French left and centre had thus been driven back in confusion and their columns separated, when, darkness coming on, further operations were suspended.

On the 11th Hope, on the left, forded the Nivelle above St. Jean de Luz, and marched on Bidart. Beresford moved on Arbonne, and Hill communicated with him on his right, while his own right was on the Atchulegue ridge and his centre faced Cambo. Soult also took up a new line of fortified camps, and,

* For a fuller account of these most interesting operations I must refer readers to Napier's "Peninsula War."

Sir J. Hope, who was ordered to make a false attack on the extreme left, had taken the enemy's works of Sans Culotte and Uroque in front of St. Jean de Luz. Here he occupied Reille's and Velatte's Divisions, threatening the camp at Belchena and the bridge near Ascain.

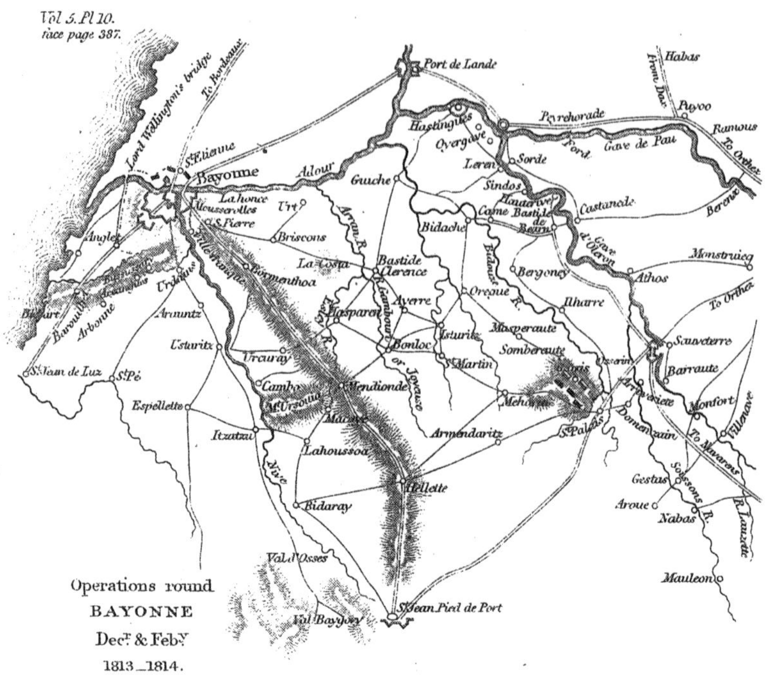
Vol 5. Pl 10.
face page 387.
Operations round
BAYONNE
Decr. & Feby.
1813_1814.

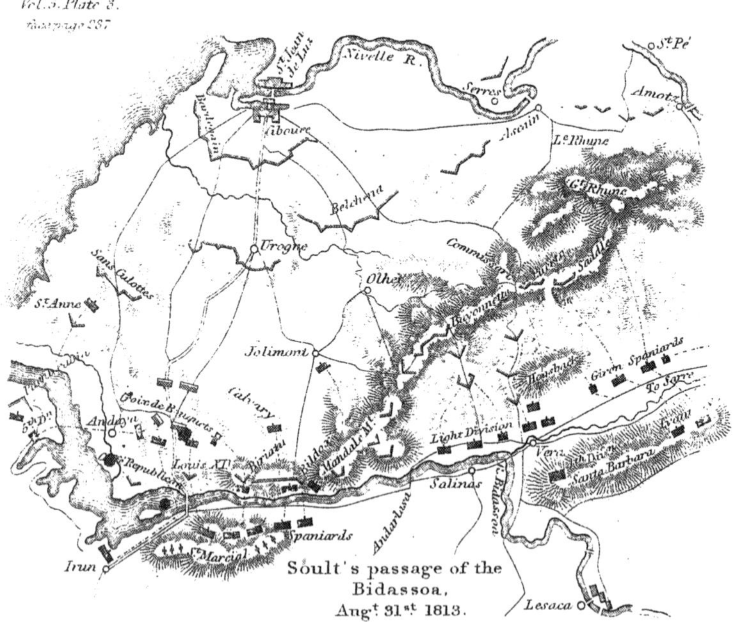
Soult's passage of the
Bidassoa,
Augt. 31st. 1813.

again retiring on the following day, took up a strong position on the ridge of Beyris, in front of Bayonne, his right resting on Anglet, his left on the fortified camp of Bayonne.

On this date also (12th) Hill was directed to menace the fortified bridge head at Cambo, and to endeavour to repair the broken bridge at Ustaritz, but owing to the heavy rain the river was unfordable, and both operations failed. The attack on the Cambo bridge was, however, renewed on the 14th,* when the 50th Regiment having entered Cambo, the enemy blew up the bridge and abandoned the tête-du-pont. The destruction of this bridge having now rendered our right flank secure, and the heavy rain having made the roads impracticable, the troops went into cantonments.

By the above operations Soult had been driven from his strong mountain position, which he had been fortifying for three months, with a loss of 4,265 officers and men, including 1,200 to 1,400 prisoners. One general had been killed, all his field magazines at St. Jean de Luz and Espellette were captured, and 51 pieces of artillery were taken.

The allies had 2 generals, Kempt and Byng, wounded, and 2,694 officers and men killed and wounded.

* The 14th is the date given in the details in a paper at the War Office. Napier gives the date as the 16th.

In December Wellington determined to force the passage of the Nive, a navigable river uniting with the Adour at Bayonne.

Hill, with the 2nd Division, Hamilton's Portuguese, some cavalry, and 14 guns, was ordered to ford that river at Cambo, and near Lahousoa; while Beresford with the 3rd and 6th Divisions was ordered to throw a pontoon bridge over, and cross at Ustaritz, on Hill's left.

In accordance with these directions, at daybreak on the 9th December, Hill, covered by the fire of his artillery, gave orders to force the passage in three columns above, two below Cambo; and the 50th Regiment, under Brevet-Major Gordon,* led the advance and forded the river breast high in a rapid current, under a heavy fire of musketry, routing the enemy on the opposite bank, though the French were strongly posted and the fords were so deep that several horsemen were drowned. The bridge at Cambo was then speedily repaired; and Hill, leaving a brigade to cover it, moved with the rest of his troops to Lormenthoa to co-operate with Beresford. Here he was joined by the 6th Division, the 3rd Division being left to cover the bridge of Ustaritz. A Portuguese brigade captured Villefranque after a sharp fight, and a brigade of the 2nd Division connected this with the rest of the division. Simul-

* Major Gordon was promoted for this service to a lieutenant-colonelcy.

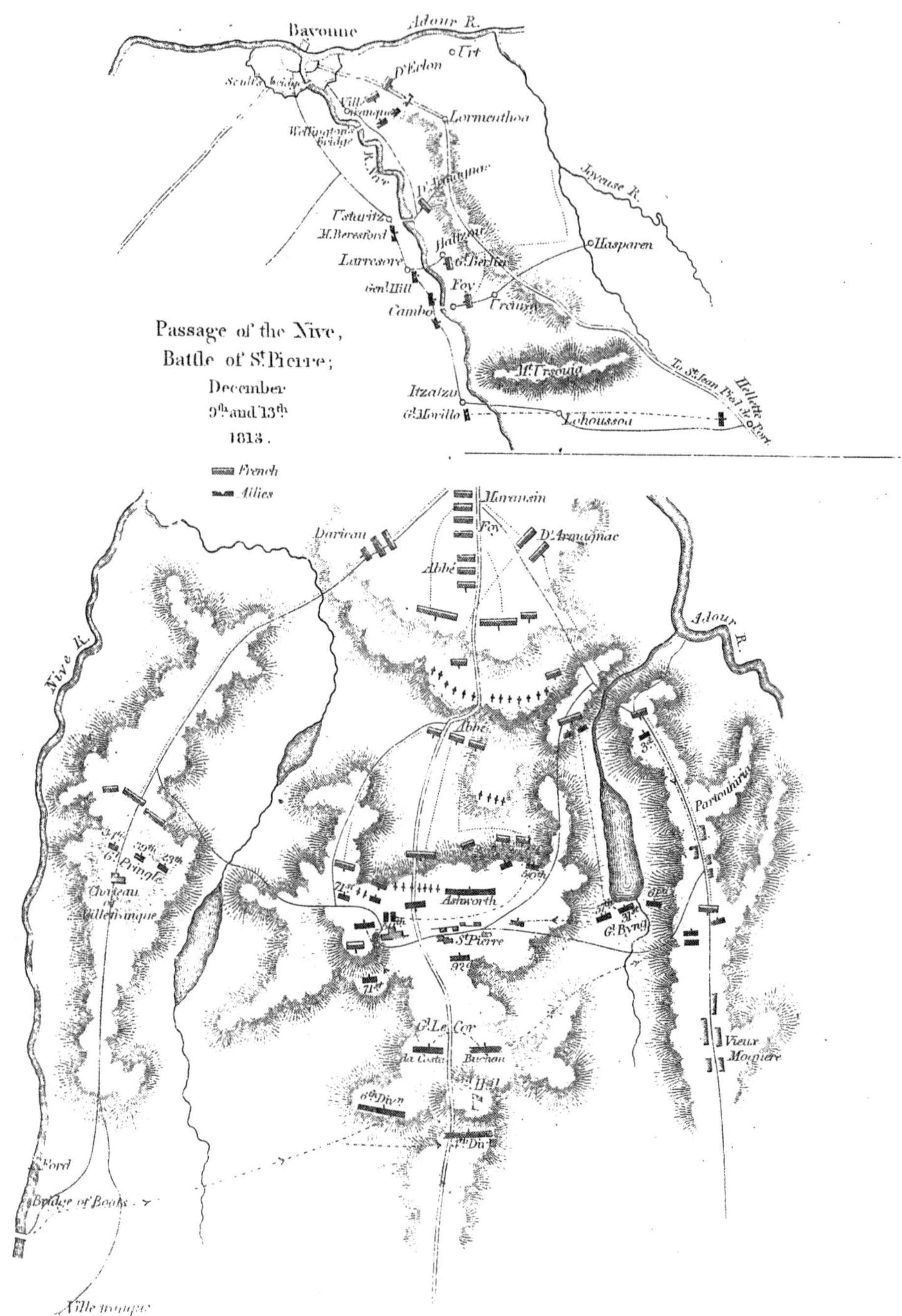
Bayonne
Adour R.
Urt
D'Erlon
Lormenthoa
Wellington's bridge
Joyeuse R.
Ustaritz
M. Beresford
Hasparen
Larressore
Gen. Hill
Cambo
Foy
Mt. Ursouia
Passage of the Nive,
Battle of St. Pierre;
December
9th and 13th
1813.
French
Allies
Itzatzu
Gl. Morillo
Lohoussoa
To St. Jean Pied de Port
Maransin
Darican
Foy
D'Armagnac
Abbé
Nive R.
Adour R.
Partouhiria
Pringle
Chateau of Villefranque
Ashworth
St. Pierre
Gl. Byng
Gl. Le Cor
da Costa
Buchan
Vieux Mouguere
6th Divn
Ford
Bridge of Boats

taneously Hope had advanced the left wing threatening Bayonne. Thus the two wings of our army were separated by a navigable river; and Soult seized the opportunity to act against one wing with his united force. With this view D'Erlon's force was brought across the river; and at dawn on the 10th a strong force was concentrated to attack Hope's wing, which was scattered in cantonments, owing to the heavy rain of the previous night. But though greatly outnumbered, the left wing maintained their position against repeated attacks, until Wellington brought the 3rd and 6th Divisions across the river. Meantime Wellington, who had been on the other side of the river when the attack began, observing that the enemy had abandoned the heights in front of Hill's Division, ordered him to occupy them.

An attack was made on Hope's left the following day (11th), and some cavalry were sent across the river to check Hill. But on the 12th heavy firing only, continued between the opposing forces on the left bank.

Soult now seeing he could not repulse our left wing, brought 7 divisions across the river to attack our right. Hill's position there occupied about two miles. On his left Pringle's Brigade occupied a wooded and broken ridge, crowned by the chateau of Villefranque, and covering the new bridge that Wellington had thrown across the river there. The right was held by Byng on the Partouhiria ridge, a strong position having a small mill-pond which nearly filled the valley in its front.

The centre was held by the 1st Brigade under Stewart. It occupied a crescent-shaped height broken by rocks and brushwood on the left, but the ground in front of the right was covered by hedges, one of which, 100 yards in front, was impracticable.* Ashworth's Portuguese and 12 guns occupied the centre in advance, behind which, on the left of the road from St. Pierre, were the 71st and 50th, the 92nd being behind that village.

One mile in front of this position rose the line of heights occupied by the French, having a broad open basin between them.

On the night of the 12th the river, flooded by heavy rain, carried away the bridge on our left.

On the morning of the 13th, under cover of a thick mist, Soult advanced against our position with 7 divisions, while an 8th, reinforced by cavalry, threatened our rear.

Thus 35,000 of the enemy attacked Hill, who had less than 14,000 men, and was partly isolated by the carrying away of the bridge on his left.

Daricau's column was directed against our left and D'Armagnac's against our right, whilst Abbé attacked our centre with great vigour; and so rapidly

* This hedge being impracticable proved of great value. If it had been practicable the 50th could not have held the wood at the critical period of this fight.

did the latter win ground on our right of Ashworth's Portuguese, that he gained a small wood on their flank; and the 71st, the right wing of the 50th, and 2 guns were sent to their aid—the 71st reinforcing the Portuguese, while the wing of the 50th recaptured the wood. This, however, weakened our centre at the moment that Abbé succeeded in gaining the summit, compelling the Portuguese and the other half of the 50th who were supporting them to retire. Then General Barnes advanced with the 92nd Regiment, and charged into the French column, which gave way.

Soult now advanced a battery of Horse Artillery, and under cover of a very heavy artillery fire, brought up his second column of attack, before which the Portuguese guns, and the 92nd retired behind St. Pierre. At this critical moment the 71st also retired, through an error of their colonel. Thus, the French had gained the centre of our position.* Ashworth's Portuguese were shattered to atoms. Generals Stewart and Barnes and most of their staff were wounded, and matters seemed desperate, yet the 50th still held the small wood on our right front, where they repulsed the repeated attacks made on them.

* Referring to the important moment when our centre was broken, Sir W. Napier quotes the following from Pringle, R.E.: "How desperately did the 50th and Portuguese fight to give time for the 92nd to rally and re-form behind St. Pierre."

Sir W. Napier says: "The right wing of the 50th and Ashworth's Cacadores never lost the small wood in our front, upholding the fight there and towards the high road with such unflinching courage that the 92nd had time to re-form behind St. Pierre. Then Colonel Cameron led it forward against the enemy's centre, colours flying and music playing, and General Hill, who had witnessed the disastrous retreat of the 71st at once descended from his post of observation, bringing that regiment forward again on our left, leading the attack himself, while Stewart brought forward the Portuguese reserves; and, on our right, the 50th, changing from defence to attack, charged into and routed a strong column of the enemy, which had advanced with great intrepidity, to within fifteen paces of them." *

Then, the enemy's column, unable to withstand this new conformation, was overthrown. Pringle's column on our left had also been successful in repulsing the attack after desperate fighting, while the assault on our right, though at first partly successful (one of D'Armagnac's brigades having actually got in their rear), was eventually repulsed. Thus Hill was successful at all points, at the moment that Welling-

* I can find no record of the left wing of the 50th Regiment, after the retreat of Ashworth's Portuguese; but they probably joined the right half battalion in holding the wood on our right front, as both the War Office Records and the Orderly Room Records, speak of the regiment there as a whole and not as a half battalion.

ton, having had the bridge of boats repaired, brought the 6th, 4th, and 3rd Divisions across the river in support, and, at once taking the offensive, drove the enemy back to their heights!

Sir W. Napier says: "It is agreed by French and English that the battle of St. Pierre was one of the most desperate of the whole war. Wellington said he had never seen a field so thickly strewn with dead; nor can the vigour of the combatants be well denied, where 5,000 combatants were killed or wounded in three hours, on a space of one mile square."

Captain North, and Lieutenants Keddle and Plunket of the 50th Regiment, were wounded, the latter severely, when charging at the head of the grenadiers; and the loss of the regiment in killed and wounded was severe.

Very wet weather and bad roads, rendered the movement of troops almost impracticable for a time; but Wellington was anxious for many reasons to advance beyond the Adour—among others, to enable him to invest Bayonne, and to give him the command of a fertile country in France, on the good roads of which his cavalry and artillery could move with greater facility.

With this view, he determined to throw a bridge over the Adour between Bayonne and the sea—an undertaking of such unexampled difficulty,* that the

* For the difficulties of construction of this great bridge,

enemy had taken no steps to prevent it. But as, in the face of serious opposition, this would have been impossible, he determined, at the same time to attack Soult's left, with the double object of preventing him from opposing the formation of the bridge, and giving himself another means of crossing the Adour, should the attempt at formation of this great bridge prove unsuccessful.

At this period, Soult, leaving 14,000 men in Bayonne under D'Erlon, had, with the remainder of his army,* taken up a very strong position with his right resting on the Adour, on the opposite bank of which Foy was in possession of the fortified position of Peyrehorade. Soult's position stretched from thence towards St. Jean Pied de Port, having the rivers Joyeuse, Bidouze, and Soisons in his front, and a line of retreat behind the broad Gaves of Oleron and Pau, branches of the Adour. (See map, page 173.)

Toward the middle of February severe frost rendered the roads practicable; and on the 12th and 13th of that month Hill's Division, which had remained watching the Mousserolles face of Bayonne, was relieved, and took post about Urcurray and Hasparen. On the 14th, Hill marched in two

see Napier's "History of the War in the Peninsula," vol. vi. p. 78, Chandos Classics edition.

* Soult's army had been weakened by the withdrawal of several divisions.

columns, the left one to turn the Joyeuse river at Bonloc, the right to dislodge the French under Harispe, from the great road leading to St. Jean Pied de Port. Both these operations were successful, and St. Jean Pied de Port was invested. (See map at p. 173.)

On the 15th, leaving the 57th Regiment at Hellette on the above-mentioned great road, Hill moved with his united columns, through Meharin upon the Garris mountain, to which strong position Harispe had retired; and the latter had hardly taken up his position, when Hill drove in his rearguard, and took up a position on an opposite and parallel ridge. Just then Wellington arrived, and his expression, "You must take the hill before dark," was heard, and repeated by the colonel of the 39th, by whom, supported by the 28th, that duty was successfully performed; the rest of the division then advancing Harispe again retired. On the 16th, Hill crossed the Bidouze at St. Palais, the cavalry and artillery by the bridge at that place, which had been repaired, and the infantry by the fords. The following day Hill marched at 8 o'clock in the morning by Domenzain, on the Soisons river, and in conjunction with the 3rd Division moved on Arriveriette, the bridge at which place was seized, and the French retired on the 18th.

During the last three days of the above operations, the 50th Regiment was engaged in several skirmishes, in one of which Lieutenant and Adjutant John Myles was mortally wounded.

Hill now took possession of all the villages along the great road from Navarens to Sauveterre, and cannonaded the double bridge head over the Gave d'Oleron at the latter place. On the 24th, the other divisions having been brought up, Wellington determined to force the passage of this Gave. With this view, the Light Division crossed the river Soisons, and forded the Gave d'Oleron at Monfort above Sauveterre without opposition; and Hill watching Navarens with Morillo's Spaniards, advanced with the 2nd Division, 3 battalions of artillery, some cavalry, and pontoons, to the ford of Villenave, which he also passed without opposition, with only the loss of 2 men drowned.

The cold was intense, the river was so deep and rapid, and the ford was so narrow, that his division was not over before dark. In connection with this movement, the 3rd Division was ordered to make a false attack on the bridge at Sauveterre, while the 7th Division threatened Peyrehorade.

Soult now ordered a retreat on the strong fortified position of Orthes; and on the 25th, the 2nd, 6th, and Light Divisions, Hamilton's Portuguese, and some cavalry and artillery, were massed in front of that place, and the bridge over the Gave de Pau at Berenx was repaired.

On the 26th Beresford finding that the works

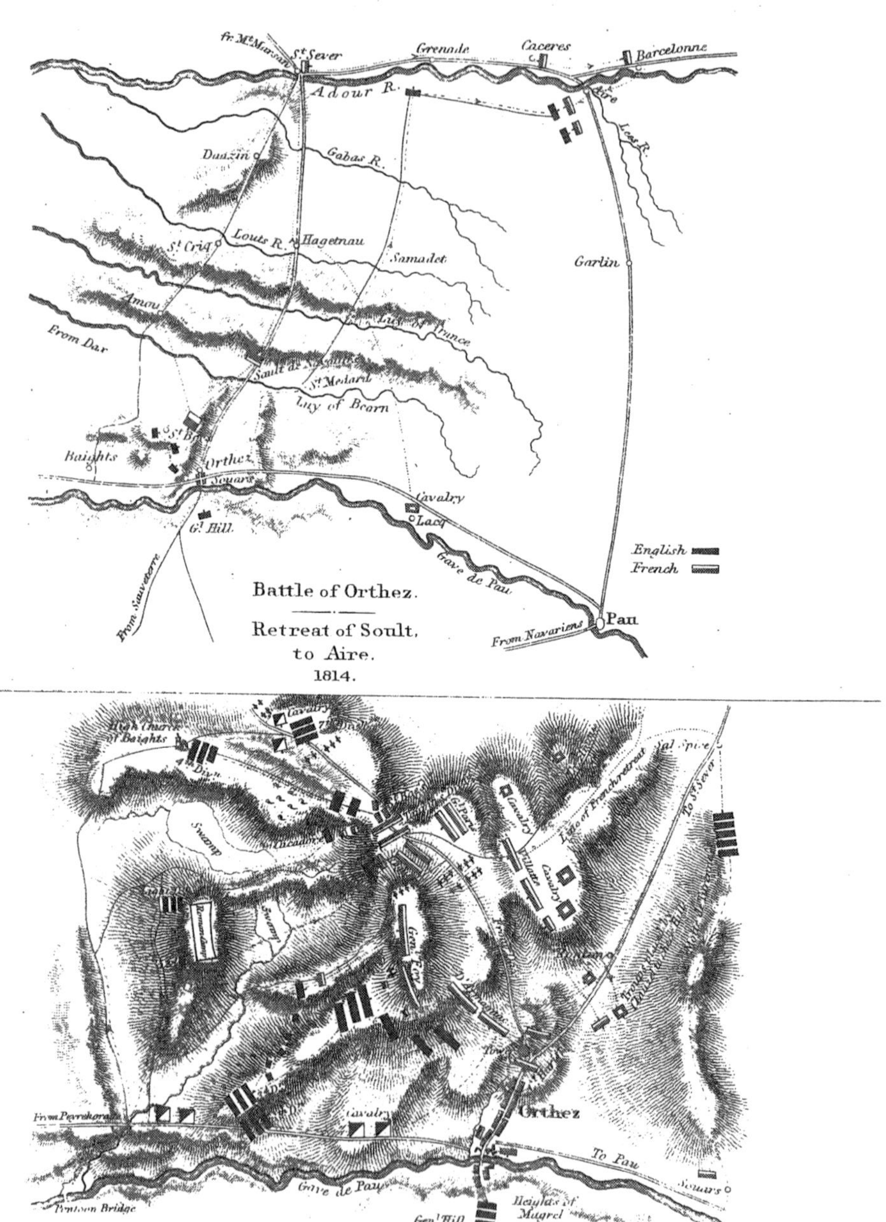
St Sever
Grenade
Caceres
Barcelonne
Adour R.
Aire
Dauzin
Gabas R.
Louts R.
Hagetnau
Samadet
Garlin
Amou
Luy of France
From Dax
Sault de Navailles
St Medard
Luy of Bearn
Baights
Orthez
Souars
Cavalry
Lacq
Gl Hill
Gave de Pau
English
French
Pau
From Navarrens
From Sauveterre
Battle of Orthez.
Retreat of Soult,
to Aire.
1814.
High Church of Baights
Cavalry
Swamp
Cavalry
Line of French retreat
To St Sever
Orthez
From Peyrehorade
Cavalry
To Pau
Souars
Gave de Pau
Pontoon Bridge
Heights of Mugrel
Genl Hill

at Peyrehorade had been abandoned, passed the river partly by fords and partly by pontoons, and took possession; and thereupon the 3rd Division passed at Berenx, and all the British columns converged on Orthes. On this date also Hope had completed his bridge over the Lower Adour, and Bayonne was invested; and the Port de Land having been seized, he was now in connection with Wellington.

At dawn on the 27th Picton's Division occupied a wooded height about cannon-shot from the French left of Orthes, covering the road from Peyrehorada; and Beresford occupied the ridge of Baights, with the 7th Division on his left, on the Dax road, having St. Boes in his front. There was a mile and a half of ground between Beresford and Picton, about the centre of which rose the height called the Roman Camp, nearly opposite to the French centre, which having a swamp in front of it seemed to forbid attack.

Wellington immediately moved the Light and 6th Divisions across the river, by a pontoon bridge that had been constructed during the night; the latter division supporting Picton on his right, while the former occupied the Roman Camp, and connected him with Beresford.

Hill remained opposite the bridge of Orthes. The 3rd and 6th Divisions won without difficulty, the lower part of the ridges opposite them; but the real attack was at St. Boes. There, with desperate gallantry, Ross's Brigade of the 4th Division, with Vasconcelli's Portuguese, broke through the scattered houses in front of St. Boes, only to be crushed by Taupin's troops massed in front, by a heavy fire of skirmishers on each flank, and by a murderous fire from the heavy guns massed on the hill, in the centre of Soult's position, and on the Dax road in rear. For three hours, they strove in vain to break through, to the more open ground beyond. Five times was the attack repeated, and as many times repulsed, with great slaughter, while Ross was dangerously wounded; nor was Picton more successful on the other flank, with the 3rd and 6th Divisions.

Then Wellington, supporting Ross's Brigade with Anson's, and massing troops on the Dax Road in rear, ordered Picton to attack the French left in mass, Hill to force the passage of the Gave, and the 52nd Regiment to cross the marsh and attack the hill in front. Colonel Colbourne led his regiment across under fire, sometimes sinking up to the men's middles; then, forming line beyond, they dashed up the hill, overthrowing a French regiment opposed to them, their unexpected appearance throwing Reille's forces into some confusion, while they took Taupin's masses in reverse. At the same time Picton was successful on the French left; and driving in Foy's Division, he seized the high ground in rear of Soult's position, occupied by D'Armagnac's Division, and established a battery of guns on a high knoll there, which ploughed through the French masses. Aided by these movements, the narrow neck of land in front of St. Boes was won; and the 4th and 7th Divisions, and Vivian's cavalry, with 2 batteries of artillery, were pushed through and established beyond. Simultaneously Hill (opposed at the town and bridge of Orthes by Harispe's and Villatte's Divisions under Clauset, who had a battalion guarding the ford of Souars), unable to force the bridge, crossed the river above Souars; and, driving back the troops opposed to him, seized the heights beyond, thereby cutting off the town of Orthes, and intercepting the French retreat on Pau. There Hill, learning of the success of the other attacks from the firing, advanced along the ridge in rear of Soult's position, parallel to another ridge, which was now Soult's only line of retreat, and converging on it at Sal Spice.

Both wings of Wellington's army having now joined, were forcing back the French, who were vigorously contesting every bit of ground, when the imminent danger of being cut off by Hill at Sal Spice, rendered their retreat hurried and confused. Hill, observing this, quickened his pace, till at last both sides began to run violently! Many men broke from the French ranks across the fields, towards the fords; and such a rush was made to gain the bridge of Sault de Navailles, that the whole country was covered with scattered bands. The cavalry sabred several hundred, many prisoners were taken, and thousands of conscripts threw down their arms. The pursuit only ceased at the river Bearn. Wellington himself was wounded in this affair; the 50th Regiment suffered but little loss.

That night, Soult, having rallied his troops, retreated across the Adour at St. Sever on Barcelone, leaving D'Erlon with two divisions at Caceres on the right bank, and Clauset, with Harispe's and Villatte's Divisions, at Aire, on the left bank. (See upper map, page 181.)

The next morning, February 28th, at daylight, Wellington pursued in three columns; Hill, with the right column, marching through St. Medard to Samadet, the other columns crossed the Adour at St. Sever and skirmished with D'Erlon's troops, which were driven back the next day, the 1st of March, on Barcelone; and Hill, marching from Samadet, reached the Adour River between St. Sever and Aire. That night heavy floods carried away the pontoon bridges over the Adour, and the main body on the right bank, being cut off from their supplies, halted till the bridges could be repaired.

On the 2nd Hill, who being on the left bank was not affected by the destruction of the bridges, moved on Aire in two columns,* expecting little

* Hill's force consisted of two divisions of infantry, one brigade of cavalry, and a battery of horse artillery.

opposition. But Clauset was in line of battle there, having his troops posted in a strong position on a steep ridge in front of Aire, high and wooded on his right, where it overlooked the river, but merging on his left into a wide tableland, over which the road to Pau led.

The two British brigades of the 2nd Division, under Stewart, were at once sent against the French right, and the Portuguese against the centre. Stewart successfully attacked the hill on our left, driving back the enemy; but the Portuguese failed in their attack on the centre and, being vigorously charged by the French, were put to flight. The other troops under Hill being still on the march, Stewart at once detached the 50th and 92nd Regiments to aid the Portuguese; and "the vehement charge of these troops turned the stream of fight: the French were broken in turn, and thrown back on their reserves." Here Lieutenant Duncan McDonnel, 50th Regiment, was killed at the head of his company. Then Byng's Brigade having come up, the French were driven right through the town of Aire, which was entered, and taken possession of by the 1st Brigade.

Two French generals were wounded, a colonel of engineers was killed, 100 prisoners were taken, and many of the conscripts threw away their arms and returned home. All the magazines at Aire fell into our hands. Harispe, however, succeeded in crossing the river Lees and breaking down the bridge.

Soult now retreated by both banks of the river Adour towards Maubourget and Marciac.

The above operations had thus resulted, not only in compelling the French entirely to abandon Bayonne, but also in cutting them off from Bordeaux, their only line of retreat being now towards Toulouse.

Wellington took the opportunity of sending a force under Beresford to seize Bourdeaux, which he took possession of on the 12th of March, and Beresford at once returned, leaving the 7th Division to hold it. The pursuit of the enemy was necessarily delayed, till the return of his troops.

The position of the allied army concentrated round Aire and Barcelone, was divided by the river Adour. Soult endeavoured to take advantage of this, by acting against one portion. With this view he gathered his divisions at Maubourget on the 12th inst., and on the 13th endeavoured to seize the high tableland between Pau and Aire. The heads of his column pointed on Aire, intercepting the line between Viella and Garlins, held by Hill's right, and menacing his posts on the Great Lees. Hill crossed that river to support his posts; but during the night he recrossed, and occupied the strong position between Aire and Garlins, which Soult had intended to seize; and the 3rd and 6th Divisions and the heavy cavalry were sent to Hill's support.

On the 14th, Soult, having driven in the outposts,

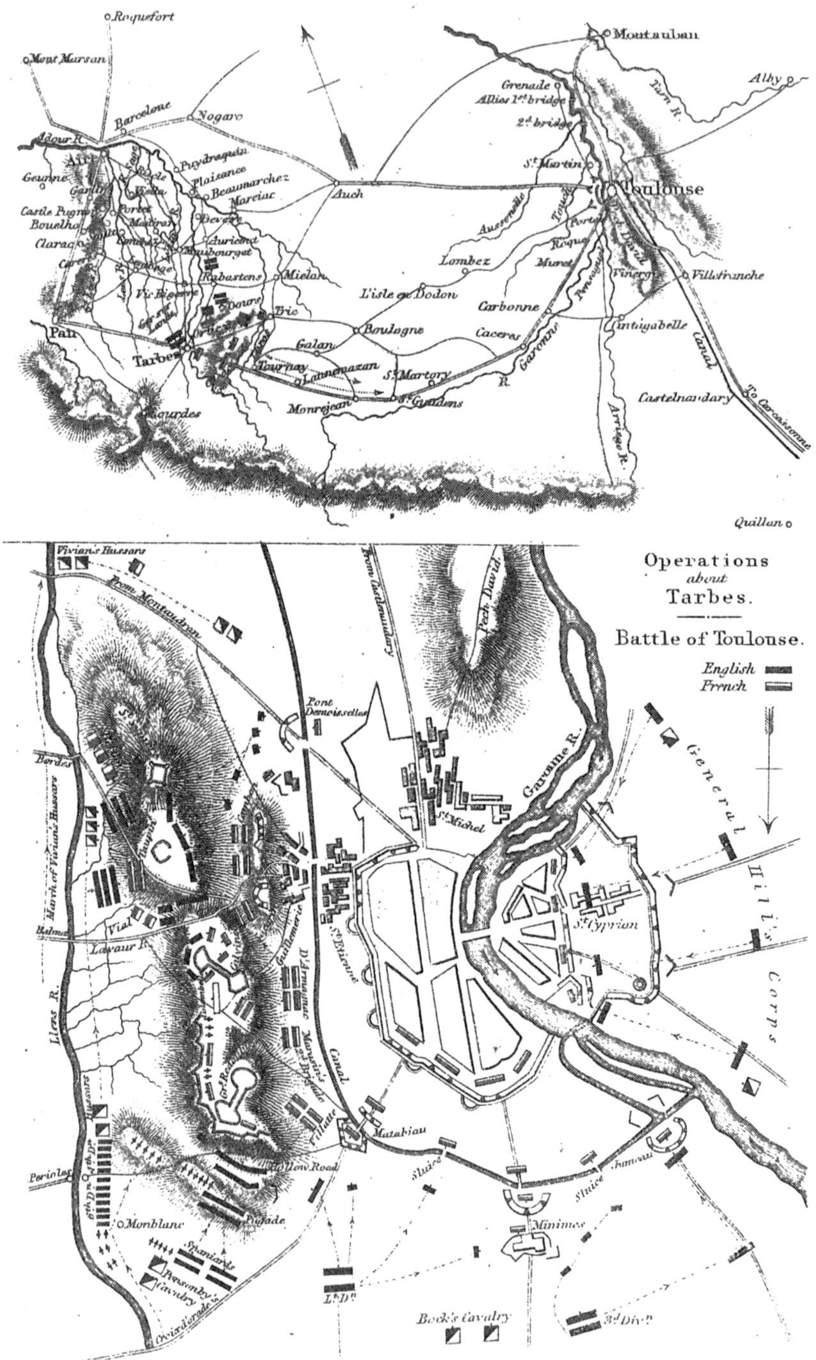
Operations about Tarbes.
Battle of Toulouse.
English
French
Toulouse
Montauban
Tarbes
Auch
General Hill's Corps
St. Cyprian
St. Michel
Garonne R.
Canal
Minimes
Matabiau
Bock's Cavalry
3d Divn
Spaniards
Ponsonby's Cavalry
Vivian's Hussars
Montblanc

examined the position, but finding it too strong did not attack; and on the 16th he commenced his retreat on Toulouse by St. Gaudens.

Wellington followed in 3 columns on the 18th. On that date the French right was turned in the valley of the Adour; and Hill, who was commanding the right column, drove back their outposts to Lembegge in a sharp skirmish, in which 80 British were killed and wounded.

Soult formed his army in line of battle on the heights of Oleac, two or three miles behind Tarbes (held by Clauset's Corps), to cover his retreat on Tourney and St. Gaudens.

At noon on that day, the Light Division was sent against his centre, assisted by the fire of Hill's artillery, while three rifle battalions attacked his left, and Hill forced the passage of the Adour at Tarbes.

The enemy's formation was broken by the three rifle battalions, and they retreated on Tournay.

The next day Soult retreated by forced marches on Monrejean and St. Gaudens; and, rapidly continuing his retreat, finally took up a position behind the Touch river, covering Toulouse. The allied army followed more leisurely, and on the 26th Beresford's Division took post behind the Aussonelle stream, facing the French army.

The position of Toulouse was exceedingly strong, the fortification being almost encircled by several rivers and a canal, while a great deal of the country around was too heavy for artillery.

At 8 p.m. on the 27th, one of Hill's brigades marched from Muret to Portel. Some men were ferried over, and a bridge over the Garonne was commenced, the remainder of the division being ordered to cross during the night; but the river proved too broad for the pontoons, and the plan was abandoned.

Wellington then drove the enemy from the Touch river on the 28th, and collected the infantry of his left and centre about Portel, Hill's Division being withdrawn to St. Roque.

During the night of the 30th a bridge was laid over the Garonne, two miles above its confluence with the Arrière river.

On the morning of the 31st, Hill crossed with two divisions of infantry, some Spaniards, artillery, and cavalry; his instructions being to seize the bridge over the Arriege at Cintagabelle, and come down the right bank to attack Toulouse from that side, while securing the ferry at Vinerge. He succeeded in crossing the river at Cintagabelle, and sent his cavalry towards Villefranche. But, finding the country too heavy for his artillery, he returned during the night, and taking up the pontoon bridge over the Garonne, left only a flying one.

The flooding of the river Garonne, prevented any further attempt to cross until the 3rd of April, when the bridge was laid at Grenade, fifteen miles on the north side of Toulouse. Owing to the river rising, it had again to be taken up, and was not finally laid till the 8th, when Wellington crossed, and, marching up both banks of the Ers river, by a brilliant cavalry skirmish seized the bridge of Croix d'Orade, which united his two columns. He then relaid his pontoon bridge at Seilh on the 9th, thus shortening his communications with Hill's Division, which remained in front of St. Cyprian, the east face of Toulouse.

The Light Division crossed the Garonne by the bridge at Seilh, at 2 a.m. on the 10th of April, and at 6 a.m. the whole army moved forward to the attack of Toulouse.

For full details of this memorable attack, I must refer my readers to the able description in Napier's work; for the war was virtually over on the 4th of April, when Napoleon signed his abdication at Fontainebleau, and I have only space to deal with the movements of Hill's Division, of which, it will be remembered, that the 50th Regiment formed part of the 1st Brigade.

Hill's orders in this attack were to menace St. Cyprian, augmenting or abating his efforts to draw the enemy's attention according to the progress of the battle on the right of the Garonne. In this position he also covered Pau. (See lower map, page 188.)

In accordance with these instructions Hill forced the first line of the St. Cyprian entrenchments and menaced the second; which, however, being more contracted and strongly fortified, could not be stormed.

This and the capture of Mount Rave,* at a cost of 8,000 of the allies, were the principal results of the fighting on the 10th.

Wellington had to pause on the 11th to renew his ammunition, and he took this opportunity of examining Hill's position at St. Cyprian.

On this date, all the light cavalry was sent on to the road leading to Carcasone, to menace Soult's retreat.

* Though Beresford's magnificent march, resulting in the capture of Mount Rave, does not come within the scope of this work, the description of this siege would be incomplete without some allusion to it—a fitting conclusion to a campaign especially noted for gallantry!

Beresford was compelled to leave his artillery behind, owing to the deep marshy country tangled with water-courses, which also impeded the infantry, to such an extent that frequent halts were required to close them up.

The Llers, an unfordable river, was on his left flank, which was under fire from the fortification of Mount Rave, and a strong force of the enemy's infantry and cavalry was on his right and in his front.

Unable to return a shot, and cruelly reduced by the enemy's incessant fire, the column steadily made its way for two miles, when they wheeled to their right, charged up hill, and took the formidable works at the summit of Mount Rave. And though unable to hold these works against the masses of men brought against them, they retained their position on Mount Rave to the end—a position which considerably influenced the evacuation of Toulouse on the following day.

Warned by the appearance of these troops that his retreat might be cut off, Soult evacuated Toulouse on the 11th, leaving behind 8 pieces of heavy artillery and 1,600 wounded men, which included 2 generals.

On the 12th Hill's troops were posted close to Baziege in pursuit, and Wellington entered Toulouse.

On the afternoon of that date, official information arrived from Paris of the abdication of Napoleon; and on the 21st a cessation of hostilities took place, which was succeeded by a general and final peace.

On the breaking up of the army, the 50th Regiment marched to Poliac *viâ* Bourdeaux, from whence they embarked for Ireland on the 20th of July, and landed at Cork on the 31st of the same month.

The following officers of the 50th Regiment were killed and wounded in the latter part of the campaign:

Killed, Lieutenant and Adjutant William Myles, Lieutenant Duncan MacDonald (killed at Aire); wounded, Captain Custance (recovered, and promoted in the 9th Regiment), Captain R. V. Lovett (died in England), Lieutenant Keddle (died at Enniskillen), Ensign Sawkins (leg amputated), Lieutenant George Bartley (recovered), Lieutenant Power (taken prisoner, placed on half-pay).

CHAPTER VII.

JAMAICA.

ORDERS were received on the 5th November, 1818, for the regiment to proceed to Jamaica, in accordance with which they left Dublin for Cork, and embarked at the latter place on the 7th and 8th of January, 1819, in the transports "John," "Fame," and "Alfred." They were detained by adverse winds till the 28th of January, 1819, on which date they sailed, arriving at Port Royal Harbour on the 7th March, and disembarking on the following day, they marched to Up Park Camp.

Their strength on landing was—

Field Officers	2
Captains	6
Subalterns	14
Staff	5
Sergeants	28
Drummers	20
Privates	608
Total	683

They were a splendid body of men when they reached Jamaica, and for four months continued in their usual health, but the season was unusually dry, hardly any rain falling till the end of July, to which is attributed the fearful mortality from yellow fever, which devastated the regiment, officers and men alike falling victims to the pestilence. Among the deaths were many officers and men, who had served with honour in the Peninsula.

The fatal progress of the disease was not checked till the end of the year, when more than half of the men, women, and children had been laid in the grave.

The casualties from the disembarkation of the regiment on the 8th of March to the following December were as follows:

Colonel Hill, C.B., Commanding,*
Captain and Brevet-Major Rowe,
Captain and Brevet-Major Montgomery,
Lieutenant Richardson,
Lieutenant North,
Ensign Barlow,
Ensign Edwards,
Ensign Harley,
Paymaster Montgomery,

* Colonel Charles Hill, C.B., who was wounded at Vimiero and who commanded the regiment at Vittoria, was still suffering from the effects of wounds, and had been advised to remain in England, but would not leave the regiment in which he had served so long.

Lieutenant and Adjutant Lyon,
Assistant-Surgeon Brown,

and 255 sergeants, drummers, and privates, besides women and children.

In order to check the mortality the regiment was divided, part being sent to Fort Augusta, and the remainder being quartered on board the "Seraphis," in Port Royal Harbour.

Toward the end of the year the sickness began to abate, and in the spring of 1820 it almost disappeared, no officers and only 58 men dying up to the end of February, 1821; but the first six months of that year resembled the year 1819 in its extreme dryness, and its fatal consequences.

In January, 1821, three draughts were sent from England to complete the strength of the regiment, consisting together of 3 captains, 1 lieutenant, 1 ensign, 1 adjutant, 1 assistant-surgeon, and 245 privates; and it was among the unseasoned men of these draughts that the fever, which again became virulent about the end of July, 1821, found its principal victims.

The deaths from the end of February, 1821, to the end of March, 1822, were—

Major E. J. Poe, Commanding,
Captain and Brevet-Major W. Masson,
Captain H. F. Jauney,
Lieutenant R. Seward,

Ensign G. Ross,
Surgeon P. Jones,
and 109 non-commissioned officers and privates.

In February of this year the regiment was distributed as follows: 2 companies at Port Royal, head-quarters and 3 companies at Spanish Town, 1 company at Fort Augusta, and 2 companies at Up Park Camp till March 6th, when they joined head-quarters.

On the 17th of April, one company under Captain Custance embarked for Savanna-la-Mar, but having lost one-third of its numbers in four months it was sent to Lucea.

The head-quarters companies occupied Spanish Town for nine months, during which the loss including detachment was: Brevet-Major Edward Scott (died on passage home), 2 lieutenants, 1 sergeant, 3 corporals, 1 drummer, and 75 privates, women, and children.

The head-quarters companies marched to Fort Augusta on the 27th of November, 1822.

Drafts were received in January and May, 1823, consisting of 1 major, 2 captains, 1 lieutenant, 2 ensigns, and 115 privates.

During this year there were only 25 deaths, and on the 2nd of November, the regiment was again moved to Up Park Camp.

A draft of 20 privates was received in February, 1824, and in November, of that year one company was sent to Port Maria, where men, women, and children were again attacked with yellow fever, and before it was withdrawn in the February following (1825) it had lost half its numbers.

The head-quarter companies, under Lieutenant-Colonel Wodehouse, marched into Spanish Town on the 13th January, 1825. They continued very healthy till June, when the fever again set in, and a further great mortality was experienced.

The men were placed under canvas outside the town for a month, but this having no effect they were moved back to the barracks, and many were sent to Port Royal for change of air.*

The head-quarters moved to Stony Hill on the 30th September, 1825, and to Up Park Camp on the 20th December, 1826. The regiment embarked for England at Kingston, under the command of Major Custance, in January and February, 1827, and landed at Gosport.

In June of that year they moved to Portsmouth, and on the 8th of August H.R.H. the Duke of Clarence, accompanied by the Duchess, presented new colours to the 50th Regiment in presence of the whole garrison. His Royal Highness, in a most complimentary address, alluded to the distinguished services of the regiment, since it was first raised by

* Orders were received in July, 1825, for a redistribution of the regiment, six companies to remain abroad and four to form a depôt in England.

his grandfather, the Duke of Cumberland, in 1756,* enumerating its badges of distinction, and paying the highest compliment to its appearance and conduct. (O.R.R.)

In consequence of this, a letter was received on the 25th September, 1827 (see Appendix), authorising the regiment to be called the 50th (or the Duke of Clarence's) Regiment, in place of the West Kent.

It will be interesting here to note that the 50th Regiment was ordered to take the title of the West Kent Regiment by Horse Guards letter of 31st August, 1782 (see Appendix), at which period regiments of infantry were ordered to take county titles in order to facilitate recruiting.

The Duke of Clarence came to the throne as William IV., on the death of his brother, George IV., in June, 1830, and on the 22nd of January, 1831 (see Horse Guards letter in Appendix), the 50th Regiment was authorised to assume the title of the

* The Duke of Cumberland, who commanded the English at the battle of Culloden, was the brother of the Duke of Clarence's grandfather, who probably spoke of the regiment being raised by his grandfather's brother, the latter word having been omitted in report.

The Duke of Clarence was the third son of George III., who was descended from Frederic, Prince of Wales, eldest son of George II., who died before his father. The Duke of Cumberland was the second son of George II. It is interesting to note that the Duke of Cumberland was instrumental in the formation of the regiment.

50th (or the Queen's Own)* Regiment, instead of that of the Duke of Clarence's Regiment, and the facings were changed from black to blue.

Thus it is to Queen Adelaide, the wife of William IV., that the 50th Regiment is indebted for the above title.

Detachments of the regiment were sent to New South Wales in 1833, and the regiment, under the command of Lieutenant-Colonel Wodehouse, landed in that colony on the 21st of November, 1834. Three companies were sent to Tasmania and two

* A difference of opinion has existed for some time, whether this title should be (Queen's Own) or (The Queen's Own).

It will be noticed that in the above letter, which is taken from the Orderly Room Records, the words "or the" are included in the brackets, but without capitals. This has led to (Queen's Own) being generally adopted in the O. R. R., and in official correspondence. It is also adopted on the Indian monument in Canterbury Cathedral, and by Captain Patterson, 50th Regiment, in dedicating his book to H.M. Queen Adelaide.

On the other hand, a letter from Lord Fitzroy Somerset, of February 15, 1831, to officer commanding 50th Regiment on the subject of this title, calls it "The Queen's Own."

Under these circumstances, I referred the matter to the Adjutant-General's Department, and received the following reply, dated 22 February, 1895:

"I beg to acquaint you, that according to our records, the title of the 50th Regiment was changed on the 14th January, 1831, from the Duke of Clarence's Regiment to the '50th' or 'The Queen's Own.'"

I have therefore adopted the latter in the title page, but in the body of the work I have used the title sanctioned by ancient custom.

to Norfolk Island, of which Major Anderson, 50th Regiment, was appointed commandant.

Two companies of the regiment, under Captain Johnstone, embarked at Sydney for New Zealand on the 30th August, 1834, to rescue a woman and nine seamen who had been captured by the natives, which service they satisfactorily accomplished, and returned in November.

CHAPTER VIII.

CAMPAIGNS IN INDIA.

The 50th Regiment received orders to hold itself in readiness to embark for Bengal, to relieve the 49th, on the 11th December, 1839.

The head-quarters and first division of the regiment embarked at Sydney, under Major Anderson, K.H. (Colonel Wodehouse, commanding, having been invalided), in the ship "Crusader," on the 29th January 1841, and the second division under Major Ryan, in the ship "Lady M'Naughten," on the 30th of that month, leaving one company under Brevet-Major Sergeant to follow with the recruits expected from England. This last party was afterwards shipwrecked in Torres Straits, and the ship "Ferguson," which carried them, became a total wreck. They, however, eventually joined the regiment, which was stationed at the time at Chinsurah and Fort William, in the Bengal Presidency. At the former station the 50th suffered greatly from cholera, and buried 23 men and 1 woman.

In September, 1841, the regiment was armed with percussion muskets.

On the 4th October, Lieutenant-Colonel Anderson (who temporarily commanded), received an order to hold the regiment in readiness for field service at Moulmain, and it arrived there by detachments, in October and November.

The 50th Regiment formed part of the expeditionary force, under Brigadier-General Logan, collected at Moulmain, on account of an expected war with the King of Burmah. It remained there for five months, when the King of Burmah having dispersed his army, the British force was broken up, and early in April, 1842, the 50th Regiment returned to Chinsurah. Here Colonel Wodehouse rejoined, and shortly afterwards received the local rank of Major-General in India; Major Fothergill joined here also, bringing a draft of 303 recruits from England; and the "Queen's Own" now mustered very strong, but shortly after the arrival of the regiment at Chinsurah, their old enemy the cholera again attacked them, and they lost Assistant-Surgeon McBean, 5 sergeants, and 69 privates.

Major-General Wodehouse received an order for the embarkation of the regiment for Cawnpore, on the 19th July, and it embarked at that date on the Hooghly, in nearly a hundred boats. A most serious gale was encountered near Barr, in the early morning of the 1st September, accompanied by heavy rain, and a number of the boats were sunk, causing immense loss of baggage, the bank of the river being littered for miles with it. Two soldiers and one woman were drowned, and nearly the whole regiment had to land for safety, causing a delay of some days. Another storm occurred off Buxar on the 26th October, by which many boats were upset; much property was lost, and one woman was drowned. A third storm was experienced near Benares; and when the regiment landed at Cawnpore on the 15th November, it had lost in transit 4 sergeants, the drum-major, 63 privates, 2 women, and 11 children.

Brigadier-General Wodehouse having been appointed to the command of a brigade at Meerut, handed over the command of the regiment to Lieutenant-Colonel Anderson, K.H., on the 31st January, 1843.

Towards the end of the year 1843, the army of the native state of Gwalior was in open rebellion against their ruler, who was supported by the Government of India; and a force called "The Army of Exercise" was ordered to assemble under Sir Hugh Gough. The 50th Regiment was told off to the second infantry brigade of the left wing of this army, under Major-General Sir John Grey, K.C.B.

The 2nd Brigade consisted of H.M. 50th Regiment, 50th Native Infantry Regiment, 58th Native Infantry

Regiment, 9th Lancers, 11th Light Cavalry, 5th Light Cavalry, Horse Artillery, Engineers, under Brigadier-General Blackhall.*

The regiment commenced its march to Bundelkund on the 14th November, leaving behind 4 officers and about 150 invalids in charge of baggage, women, and children. After remaining in camp near Kouet for three days, it marched by easy stages to Duboi, where the brigade was completed. The brigade remained here waiting for orders until the 18th December, the head-quarters and main body of the Army of Exercise being near Agra, on the river Jumna. On that date the army moved forward, fording the river Scinde in three columns, on the 23rd of that month.

A halt was made at Somaree for a short time on the 27th. Here information was received, that the Mahratta troops were in considerable force in the Antree Pass, about eight miles in front, on the main road to Gwalior. A further advance was made on the 28th. On this date Brigadier-General Blackhall, commanding the 2nd Brigade, met with a pistol accident, and Lieutenant-Colonel Anderson assumed command, Major and Brevet Lieutenant-Colonel Petit taking command of the regiment. An engagement with the enemy being now imminent, the

* The 1st Brigade was composed as follows: H.M. 3rd Buffs, 8th Irregular Cavalry, 8th Light Cavalry, 51st Native Infantry, 39th Native Infantry.

infantry marched with loosened ammunition, and the artillery with lighted fuses.

On the 29th December Sir J. Grey's force, leaving the Antree Pass on their right, moved to their left on Himutgur and Punniar, in order to turn the right of the enemy's position.

The advanced guard of the 2nd Brigade was formed from their light companies, that of the Queen's Own leading. The brigade arrived on its camping ground about noon, after a hot march of about sixteen miles, having passed several of the enemy's forts on the way, one a very large one at the top of a high hill. Heavy firing commenced on our line of baggage about 2 o'clock, and two hours later the enemy were observed taking up a strong position on a chain of hills.

PUNNIAR.

The whole force was immediately ordered to be ready for action.

The Buffs, with the 1st Brigade, marched off first, and about 5 p.m. the 50th with the rest of the 2nd Brigade followed. For about three miles, they marched over the ground where the Buffs had been engaged, passing their dead and wounded, and a party of their men working one of the enemy's captured guns. The 2nd Brigade were then ordered to attack the enemy's left, formed on the crest of a hill, a movement which they executed in a most brilliant manner.

The following description is from the diary of the acting adjutant of the regiment:

"We passed the General and Staff, who ordered us to clear a small valley to the right, the Buffs having gone to clear one to the left. As we reached the head of the hill which overlooked the vale, we formed line on the grenadier company, and there being no room for the two native regiments (50th Native Infantry and 58th Native Infantry), they remained in open column of companies. Directly we reached the top of the hill—as well as during our progress to the top of it—the enemy's cannon-balls were falling right and left of us, but being badly directed did us no harm. We moved a few paces over the hill, when they opened a heavy fire of grape and canister upon us, with four guns planted about fifty paces from the bottom of the hill, besides a tremendous fire from their infantry, who were in a small ravine. We made the best of our way down the hill, which was very high and steep, keeping the best order possible, and continuing our firing the whole time. We halted at the bottom under cover of a small bank and hedge, keeping up our fire for about ten minutes, when we were ordered to charge, which we did with a glorious cheer. But so well did the enemy stick to their guns, that the last discharge took place when we were within ten yards of them, and the gunners were only driven from their guns at the point of the bayonet. So determined were they, indeed, that until actually unable to move from wounds, they cut away with their sharp sabres at our men, many of whom were severely wounded by them. Thus ended this short but sharp skirmish, with the capture of four guns (one a large brass one) and a few prisoners. It was 8 o'clock when we ceased firing, had it been light we should have done more execution. We collected our killed and wounded and marched home, and the regiment was under arms all night."

In the above action Lieutenant-Colonel Anderson, 50th Regiment, commanding the brigade, fell severely wounded. Captain Cobham was killed at the head of his company, and Lieutenant-Colonel Petit, commanding the regiment, had his horse shot under him while leading the charge. Besides this 8 privates were killed, and 3 sergeants, 1 drummer, and 28 privates were severely wounded.*

Captain Cobham and the private soldiers who were killed were buried that evening, the General and all his Staff attending.

The following general order was issued next day:

"The Major-General begs to congratulate the officers and men, engaged with the enemy yesterday,

* The Orderly Room Records give only 20 privates wounded. I take the larger number of 28 from the figures of the acting adjutant of the regiment at the time (Lieutenant Bellars).

on the brilliant victory they have gained over a force five times their number, occupying a very strong position, defended by all their artillery and infantry. The former stood until cut down at their guns.

"No troops could have behaved more gallantly or fought under greater disadvantages, having come off a long and fatiguing march of sixteen miles.

"The Major-General is convinced that had another hour's daylight remained, and the cavalry been brought into play, the whole of the enemy's army would have been immolated. The Major-General will not fail to lay before the proper authorities, the names of those officers and soldiers, who conspicuously distinguished themselves.*

"The Major-General laments that the victory should have been so dearly bought, at the loss of so many valuable lives."

On the morning of the same day (December 29th), an important victory was gained at Maharajpore, by the right wing of the army under Sir Hugh Gough, G.C.B. Major Ryan of the 50th Regiment was present, attached to the 39th Regiment.

* Extract from the despatch of Sir J. Grey, 30—12—43 "Major Petit, commanding H.M. 50th Regiment, distinguished himself by the gallant charge he made down hill at the head of his regiment. I have to bring to your Excellency's notice the able assistance I derived from Captain G——, D.A.A.G., and also from my A.D.C. Captain Tudor, H.M. 50th Regiment."

Extract from General Order by the Right Honourable the Governor-General of India: "H.M. 3rd Buffs and 50th Regiment added new lustre to the reputation they gained in the Peninsular War."

On the 4th January, 1844, the two wings of the army formed a junction at Gwalior, and took up a position within three miles of that fortress.*

In consequence of the above victories, the authority of the Maharajah was restored, and the Mahratta army was broken up.

The 50th Regiment left Gwalior on the 25th January, and returned to their old station at Cawnpore, which they reached on the 14th February.

A detachment of the 50th under Captain Stapleton, on board the "Runnymede," was wrecked on the Andaman Islands on the 11th March.†

Owing to the unhealthiness of the season the "Queen's Own" suffered a great deal from cholera and ophthalmia at Cawnpore, losing 119 men and several officers. In consequence of this they were ordered to Loodiana, and commenced their march on the 15th October, arriving at their destination on the 7th December.

Lieutenant-Colonel Ryan, who had been in charge of the convalescent depôt at Landour, arrived on the 26th October, and took over the command of the regiment, Lieutenant-Colonel Anderson having been invalided.

* Gwalior was taken by a party under Lieutenant-Colonel White on the 4th February, and totally evacuated on the 5th.

† Captain Stapleton was in command of the detachment of the 10th and 50th Regiments. His conduct on that occasion, in conjunction with the officer commanding the detachment of the 80th Regiment, elicited the strong approbation of the Commander-in-Chief.

Further hostilities appeared to be imminent in November, 1845, as the Sikh army was reported to be making preparations to cross the Sutlej, at a time of profound peace, with the avowed intention of attacking Ferozepore.

The 1st Division of the Sikh army crossed the Sutlej river on the 13th December, and the whole army was encamped near Ferozepore on the 16th.

The 50th Regiment was inspected by the Governor-General, Sir H. Hardinge, on the 11th December, who told them that they would soon be employed. He stopped near the colours of the regiment, and told Colonel Ryan that he hoped his men would behave as he had seen the old 50th do in the Peninsula. The Colonel replied that "they were only anxious to be tried."

Orders were read at the evening parade on the 13th, directing the regiment to hold itself in readiness to march next morning; but further orders would be issued if they were actually to proceed.

This further order was issued about 11 p.m., and sent to Colonel Ryan, who was living some distance from the camp. He sent an orderly up with it at once, but the orderly seems to have made some mistake about delivering it; and the result was the 50th did not parade at daybreak on the 14th, as it should have done, and the rest of the brigade marched off without them. They, however, started about 8 a.m., and overtook the Native Infantry

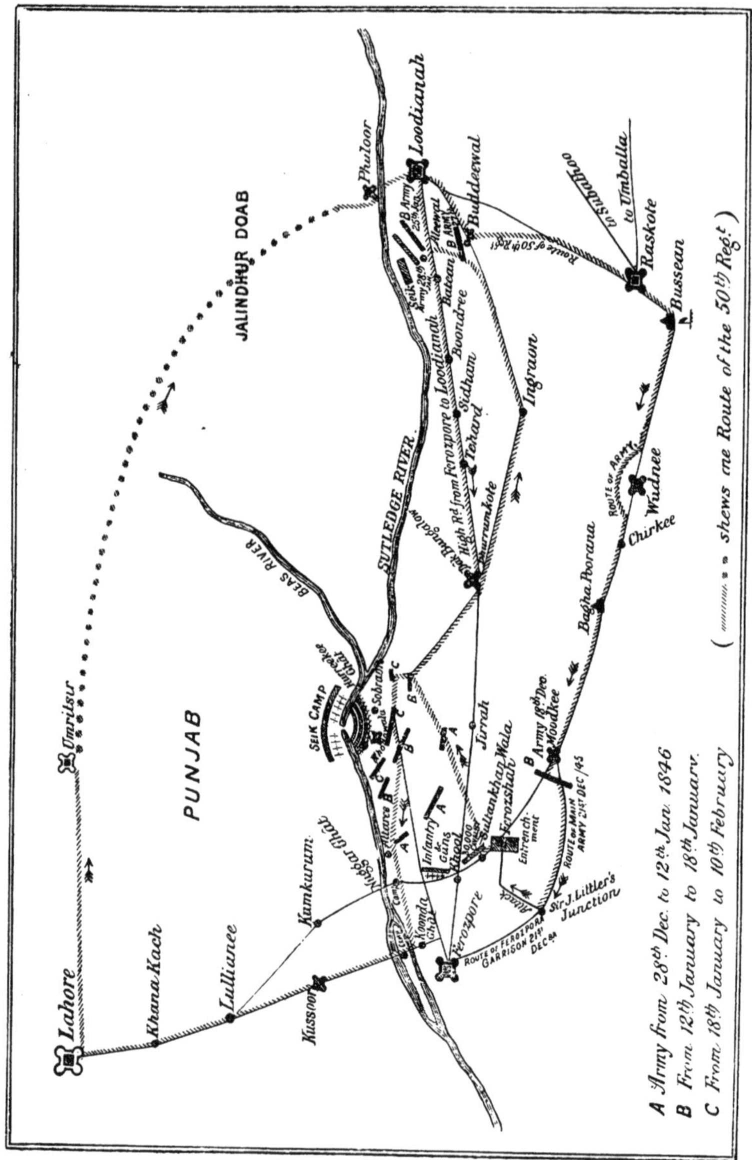
Lahore
Khana Kach
Lullianee
Kussoor
Kamkurum
Umritsur
PUNJAB
JALINDHUR DOAB
Phulloor
Loodianah
Buddeewal
Raskote
to Subathoo
to Umballa
Bussean
Ingraon
Wudnee
Chirkee
Budha Poorana
Route of Army
Moodkee
Army 18th Dec.
Ferozshah
Entrenchment
Sultankhan Wala
Jirrah
Ferozpore
Route of Ferozpore Garrison 21st Dec.
Route of Main Army 21st Dec /45
Sir J. Littler's Junction
Koonda Ghat
Nuggar Ghat
SUTLEDGE RIVER
BEAS RIVER
SEIK CAMP
Sobraon
Hurreekee Ghat
Infantry & Guns
High Rd from Ferozpore to Loodianah
Dhurumkote
Tehard
Sidham
Boondree
Batean
Route of 50th Regt
A Army from 28th Dec. to 12th Jan. 1846
B From 12th January to 18th January
C From 18th January to 10th February
(shews the Route of the 50th Regt.)

regiments of the brigade near the fort of Buddiwal, which was taken without resistance. Some companies of the Native Infantry regiments took possession of the fort and four old brass guns, which were at once sent off to Loodiana in charge of four companies of Sepoys.*

The brigade afterwards marched to Bussean, which was reached at sunset without halting, after a fatiguing march of thirty miles over sandy roads—no tents or baggage coming up until the following morning.

General Sir H. Smith and the Umballah force came into camp on the 15th, and his division was told off as follows:

1st Brigade, under Brigadier-General Hicks—H.M. 31st Regiment, 24th and 47th Native Infantry Regiments. 2nd Brigade, under Brigadier-General Wheeler—H.M. 50th Regiment, 42nd and 48th Native Infantry Regiments.

* After the capture of Buddiwal, Lieutenant Brockman, who in the hurried departure of the regiment had come without baggage, went back on foot for it. He must have been captured by the enemy, as he was never heard of afterwards. Two other officers who went back at the same time on horseback rejoined in safety, having seen nothing of him. Two British prisoners were afterwards reported by spies to be detained in the enemy's camp, and one of these it was supposed might have been Lieutenant Brockman; but nothing further was ever heard of him, and he was eventually struck off the strength of the regiment by a court of inquiry. He was reported as "killed in passing Wudnee."

Prince Waldemar of Prussia, who had been attached to the regiment for the past six weeks, accompanied them; and the doctor in attendance upon him was killed at the battle of Ferozeshah.

On the 16th December the first bugle sounded at 2 a.m., and the 2nd Brigade marched off at 3, being the leading brigade of the whole army, and the 50th the leading infantry regiment of the brigade.

The whole force reached Wudnee, seventeen miles off, about 1 p.m., and drew up in front of the fortress, which contained a strong garrison of the enemy and refused to surrender. Preparations were made for an attack, but eventually an arrangement was made by which the villagers undertook to provide provisions, and General Wheeler's force, having made a detour, encamped on the other side, leaving the reduction of the fort to the division in rear, which was accompanied by the heavy guns. "The Queen's Own" marched early next morning for Bhago Pooranah, about ten miles off.

The regiment marched about 4 a.m. on the 18th. After a short march, the enemy were reported in front, and the cavalry and artillery were sent ahead, the infantry being hurried on after them. After a fatiguing march of about seventeen miles, during which the 50th suffered a good deal from want of water, they halted near a large tank close to the village of Moodkee, the light company and No. 8 being ordered out on piquet.

About 3 p.m. the whole force was called to arms, the enemy having taken up a strong position in our front.

MOODKEE.

The action of Moodkee was commenced by cavalry charges and heavy firing from our field artillery.

The "Queens Own" Regiment was formed in quarter column, left in front, on the extreme right of the line of infantry, having the 3rd Dragoon Guards and a battery of artillery on its right front.

In this order it advanced through a country thickly covered with low jungle, and formed line just before coming within cannon range of the enemy, the artillery still maintaining their position on its right. The enemy's fire soon began to tell severely, and the nature of the ground concealed several of their skirmishers, who also posted themselves in the trees. The brigade at first formed square to resist the enemy's cavalry, but Sir H. Smith riding up, ordered the 50th to re-form line and advance, which it immediately did, closing with the Sikhs with three cheers, and after a desperate struggle, during which the enemy stood their ground, and returned volley for volley; by a final charge, the 50th succeeded in capturing the guns opposed to them, and driving back the masses of the enemy's infantry. The Native Infantry regiments remained in square for some time after the advance of the 50th, and the fire of one of them came into their ranks.

The 50th was afterwards drawn up in quarter column, in advance of the position captured from the enemy, where it remained till about 12.30 o'clock, when it returned to camp at Moodkee, two companies having been sent out to collect the wounded, who were brought in on the "Razais," or native quilts, captured from the enemy.

The loss of the regiment in the above was: Assistant-Surgeon Graysdon * and 25 privates killed, and the following wounded: Captain Needham (severely), Captain Long (slightly), Lieutenant Young (died of wounds), Lieutenant Bishop (died of wounds), Lieutenant Carter (slightly), and 94 privates.

During the advance Brigadier-General Wheeler was wounded, and Colonel Ryan took command of the brigade, Lieutenant-Colonel Petit taking command of the regiment. Both Colonels Ryan and Petit had their horses shot under them.

The troops were drawn up in front of their camp on the following morning (December 19th), the enemy being about four miles off. Towards evening, they moved off to their intrenched camp at Ferozeshah, which was about ten miles off, between Moodkee and Ferozepore, the latter place being occupied by Sir J. Littler's force.

* The bodies of Lieutenant Carey of the 29th Regiment and the Roman Catholic priest of Loodiana, who went into action with the 50th, were also brought in. The former was killed with the grenadier company, and the body of the latter was dreadfully mutilated.

Orders were given on the evening of the 20th, for two days' provisions to be cooked, and preparations to be made for an early start next morning.

The sick and wounded, with an officer from every regiment, and the 17 captured guns (which had been dismounted), were left in the fort at Moodkee.

Sir H. Smith's Division marched about 4 a.m. on the 21st, guided by three large stars until daylight.

A junction was effected with a force from Ferozepore, under Sir J. Littler, about 2.30 p.m., and preparations were at once made to attack the enemy's position.

FEROZESHAH.

Sir H. Smith's Division was at first in reserve, about 300 yards in rear of the first line, which opened with musketry and advanced on the enemy's position; but the severe fire of round and grapeshot from the enemy's works, told so heavily upon them, that the 2nd Brigade of Sir H. Smith's Division, under Lieutenant-Colonel Ryan, was ordered to advance to their support. The 3rd Light Dragoons now made some gallant charges; and the 50th Regiment, advancing rapidly with the 2nd Brigade, soon came into a gap in the first line and steadily advanced, though the enemy's grapeshot tore through their ranks, and their matchlock men kept up an incessant fire. Two staff officers fell close to them—Major Somerset, who had brought an order, near the colours of the regi-

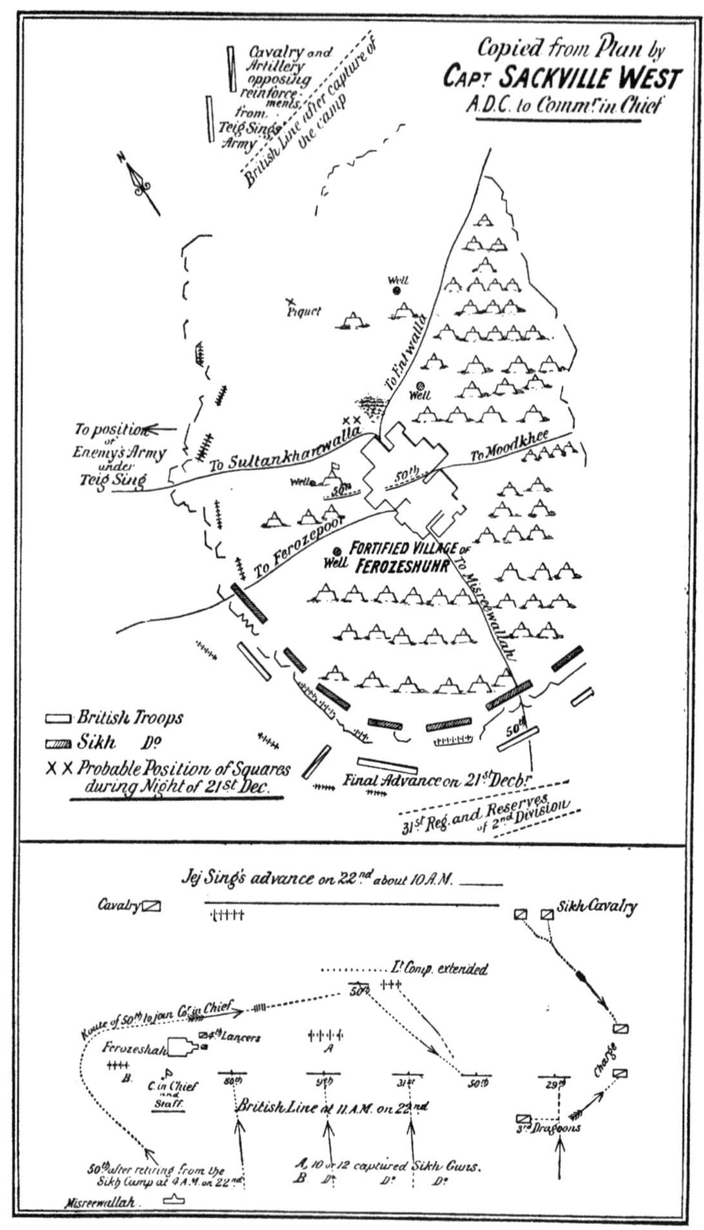
Copied from Plan by
CAPT. SACKVILLE WEST
A.D.C. to Comm.r in Chief
Cavalry and Artillery opposing reinforcements, from Teig Sing's Army
British Line after capture of the camp
N
Piquet
Well
To Ferozepoor
To Sultankhanwalla
To Moodkhee
To Misreewallah
FORTIFIED VILLAGE OF FEROZESHUHR
To position of Enemy's Army under Teig Sing
50th
British Troops
Sikh Do
X X Probable Position of Squares during Night of 21st Dec.
Final Advance on 21st Decbr.
31st Reg. and Reserves of 2nd Division
Jej Sing's advance on 22nd about 10 A.M.
Cavalry
Sikh Cavalry
Lt Comp. extended
Route of 50th to join C. in Chief
3rd Lancers
Ferozeshah
C. in Chief and Staff
British Line at 11 A.M. on 22nd
80th
9th
31st
50th
29th
Charge
3rd Dragoons
50th after retiring from the Sikh Camp at 4 A.M. on 22nd
A, 10 or 12 captured Sikh Guns.
B. Do Do Do
Misreewallah

ment, and Major Broadfoot near the right. The "Queen's Own" now brought their right shoulders up, and after a few rounds of musketry charged into the trenches,* driving the Sikhs back into their camp at the point of the bayonet, Corporal Hale, light company, and Private Johnson of the grenadiers, capturing two of the enemy's standards.†

* Sergeant J. Godwin, 50th Regiment, who was present with his company, says: "Just as our company had their feet on the entrenchment, Sir Harry rode along behind the regiment and said, 'Into them, my lads! the day is your own.'"

† A third standard was afterwards captured, and probably by Sergeant-Major Cantwell, at Sobraon.

The Orderly Room Records say: "Sergeant-Major Cantwell much distinguished himself, but was killed at the moment of victory when the regiment was entering the trenches," and a paper at the War Office adds, "and just as he had captured a standard from the enemy." Sergeant Godwin, one of the few survivors of this glorious campaign, does not remember a standard being taken. He adds, "Sergeant-Major Cantwell was found killed and stripped at the river side."

This is not incompatible, however, with the narrative of his capturing the standard, as he would have confided it to an escort; or with that of his being killed at the moment of victory, for he would have been one of the first men into the trenches, and would have followed up the retreating enemy.

These three standards for many years decorated the regimental mess room; but in course of time very little of the colours remained, and the author of this work, when in command of the Regiment, had them mounted in an air-tight case with a glass front, the frame of which is made from their own flag-staffs and ornamented with the Crown and Sphinx, and the Kentish horse in silver, with an inscription at the base.

This now hangs in All Saints' Church, Maidstone, where the old Crimean colours have also found an honourable resting place.

The 50th were the first entire regiment that succeeded in entering the enemy's works. They charged on through the camp; and the right wing, accompanied by Sir H. Smith, charged through the village of Ferozeshah, the left wing passing outside. It took some time to drive the enemy out of the village, as they took advantage of the approaching darkness, and of the small houses and narrow streets. Here parties of the 9th Regiment, 1st Europeans, and several native regiments came up. Eventually the two wings of the regiment were united outside the village, and they formed a square for the night in a position selected by Sir H. Smith, on the eastern side of the camp. Parts of other regiments formed another large square a little distance off, irregular from its being formed from many regiments, but effective.

These squares retained their position all through the night of the 21st, though harassed by the enemy's artillery and musketry, and by parties hovering round in the darkness, and firing into the squares whenever opportunity offered, while our men were hardly able to return a shot, not knowing the position of the rest of the army.

The light company of the 50th and a few parties from other regiments, were sent out a little distance in advance, but the Sikhs discovering their small number fired incessantly on them, and bringing up a gun in the darkness, fired a charge of grape into the light company, which had to retire, and eventually lost their way. Lieutenant Frampton, however, who was with them, having picked up one of the Sikh bugles, sounded the regimental call on it, which was answered from the regiment; and thus they were able to make good their retreat. The acting adjutant of the regiment (Lieutenant Bellars) in his diary thus writes of this night:—

"No one can imagine the dreadful uncertainty. A burning camp on one side of the village, mines and ammunition waggons exploding in every direction, the loud orders to extinguish the fires as the Sepoys lighted them, the volleys given should the Sikhs venture too near, the booming of the monster guns, the incessant firing of the smaller ones, the continued whistling noise of the shell, grape, and round shot, the bugles sounding, the drums beating, and the yelling of the enemy, together with the intense thirst, fatigue, and cold, and not knowing whether the rest of the army were the conquerors or conquered—all contributed to make this night awful in the extreme."

A council of war in the early morning, decided that a retreat to rejoin the rest of the army, was necessary; and about 3 a.m. on the 22nd they moved off, after gallantly carrying the batteries still left in possession of the enemy. The light company of the 50th Regiment led the way; the remainder of the regiment at first brought up the rear, but were eventually brought to the front to meet any attack, for (harassed and exhausted as the brigade now was, the artillery without ammunition, and weakened by the cavalry and some of the artillery having gone into Ferozepore by mistake), they were threatened with a new danger from the army of Teig Sing, reported to be 30,000 strong, which was covering the retreat of the Sikh army.

The 50th Regiment retreated by Misreewallah. They got some dholies *en route* from the field hospital of the 62nd for the transport of the wounded, and marched almost round the entrenched camp. (See Plan.) Heavy firing was heard as they approached the remainder of the army. They came up between the British line and the enemy, and formed in front of the centre, four British guns being on their right flank; but these, after firing three or four shots, retired.

Teig Sing's artillery now opened a heavy fire, and continued their advance.

Thirteen or fourteen of our men being wounded, the light company was thrown out in extended order, and the regiment was moved to a more favourable position behind a bank. The enemy's artillery kept up their fire for some time, the British artillery having gone to Ferozepore for ammunition. The 50th was afterwards retired into the general line, but the light company remained out for some time.

About 2 p.m. the enemy steadily retired, and the

men of the regiment, who had had no food for two days, except the rations they carried in their haversacks, managed to get a small bullock, which was distributed as rations. They slept on the ground that night, in quarter distance column with piled arms.

The killed and wounded during the two days' fighting were :

Killed.

24 rank and file.

Wounded.

Captain Knowles, slightly.
Lieutenant Mouat, slightly.
Lieutenant Chambers, slightly.
Lieutenant Barnes, severely.
Lieutenant White, severely.
Lieutenant and Adjutant Mullen, severely.
7 sergeants, 2 drummers, and 80 rank and file.

The next morning, many explosions having taken place, through tumbrils and cases of powder (left by the enemy) igniting, the regiment was withdrawn 200 or 300 yards; and parties were employed all day in burying the dead, and bringing in the wounded, many of whom had lain for more than twenty-four hours without surgical assistance.

On the 24th all the wounded were sent back to Ferozepore, and the regiment paraded at 9 a.m. and marched to Sultan Khan Wallah. The baggage arrived from Moodkee on Christmas Day, on which date also a despatch was read to the regiment from the Governor-General, in which he announced "the repulse of the Sikh force near Moodkee, and the capture, on the evening of the 21st and the morning of the 22nd, of their entrenched camp, with 70 pieces of cannon, defended by 60,000 men, near the village of Pheroz-shuhur.* Upwards of 90 pieces of the enemy's artillery have been taken in these two operations."

Sir H. Smith also addressed them, and expressed his warmest approbation of their gallant conduct.

They marched to Gammonwalla on the 28th, and on the 1st of January, 1846, the division took up a position on the river Sutlej, opposite to the enemy's camp on the other side of the river.

Sir H. Smith was detached with a part of his division towards Loodiana on the 17th of January, and on the 22nd the 2nd Brigade of his Division, which had been inspected by the Governor-General on the previous day, marched to join him. Accompanied by the Governor-General's body-guard and 4 guns, they reached Durrumkote that evening and Sidham the next day, after a heavy march of fifteen miles. Here it was found that a very large force of the Sikh army under Rungeit Singh was between them and the rest of the division, and it was considered advisable to return to Durrumkote, where they arrived about 6 p.m., much exhausted by a fatiguing march of over thirty miles, through heavy sand, without even time to cook their rations.

* This place is also spelt Ferozeshur and Ferozeshah.

The following morning the brigade, having been reinforced by the 8th Light Cavalry, resumed their march by a more circuitous route, and rejoined their division at Buddewal on the 26th, the enemy having abandoned this position on the 22nd, and retreated to their entrenched camp on the river Sutlej beyond Aliwal.* Thither Sir H. Smith prepared to follow them on the 28th. The composition of his force was as follows:—

STRENGTH OF TROOPS UNDER MAJOR-GENERAL SIR H. SMITH, G.C.B.

	Officers and Men.	
Artillery—Major Lawrence:		
Horse Artillery (6-pounders) . .	22	
3rd Artillery (9-pounders) . . .	6	
Shekawattie Brigade	4	
Total Artillery	——	32
Cavalry—Lieutenant-Colonel Churton:		
1st Brigade (MacDowell)—		
16th Lancers	530	
3rd Light Cavalry.	372	
4th Irregulars	398	
	——	1,300
2nd Brigade (Stedman)—		
Bodyguard	351	
1st Cavalry	422	
5th Cavalry	402	
	——	1,175
(Major Foster)—		
Shekawattie		631
Total Cavalry		3,106

* A draft of 99 privates, with the following officers, arrived on the 27th:

Ensigns Davenay, Purcell, Slessor, James, and Farmer.

Infantry—1st Brigade (Lieutenant-Colonel Hicks)—		
31st Regiment	544	
24th Native Infantry	481	
36th Native Infantry	571	
		1,596
2nd Brigade (Lieutenant-Colonel Wheeler)—		
50th Regiment	494	
48th Native Infantry	857	
Sirmoor Ghoorkas	781	
		2,132
3rd Brigade (Lieutenant-Colonel Wilson)—		
53rd Regiment	699	
30th Native Infantry	824	
Shekawattie	625	
		2,148
4th Brigade (Lieutenant-Colonel Godby)—		
47th Native Infantry	713	
Mysore Ghoorkas	586	
		1,299
Total Infantry		7,175
Sappers		28

ALIWAL.

After marching about nine miles the enemy were seen drawn up in order of battle.* Preparations were at once made for an attack.

* Passing over very rough ground, the 50th reached the village of Pourein about 9 o'clock, and on the rising ground beyond saw the whole Sikh force drawn up, with their cavalry in front. (Extract from Lieutenant Bellars' diary.)

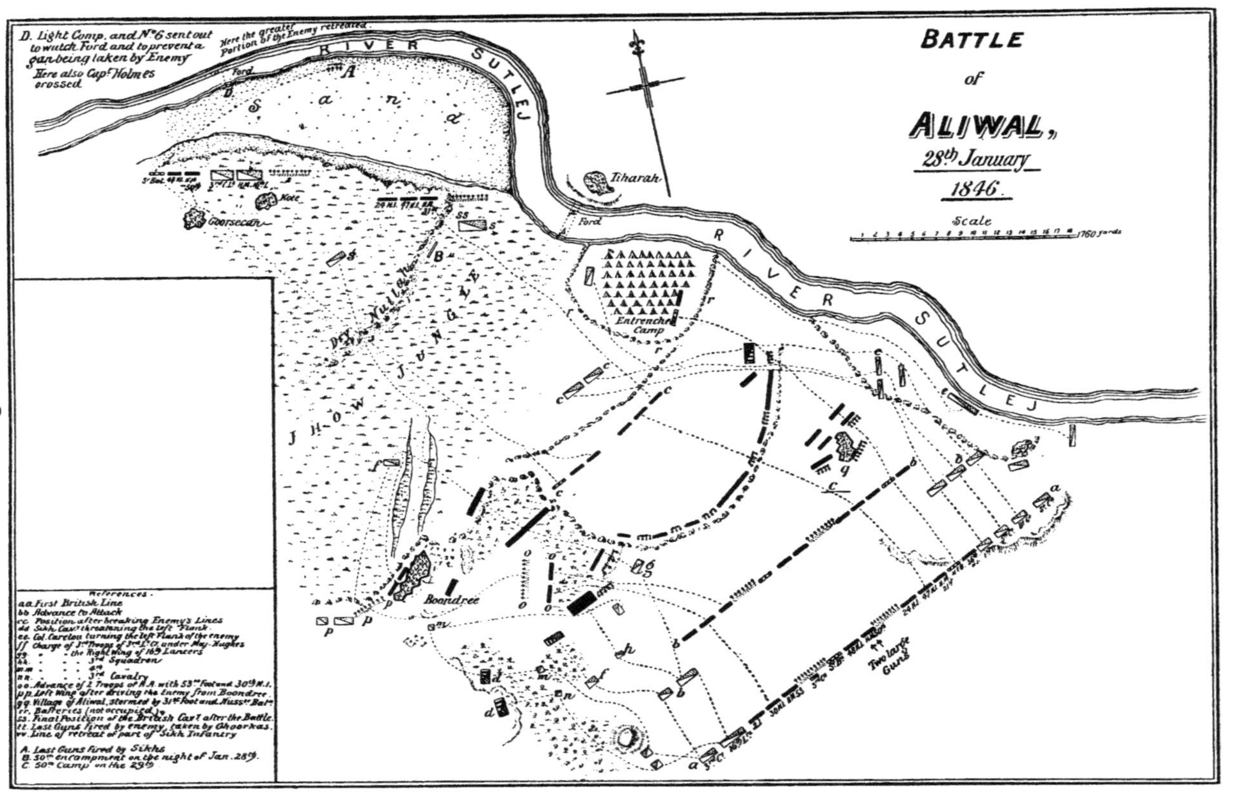
BATTLE
of
ALIWAL,
28th January
1846.
Scale
1760 yards
RIVER SUTLEJ
Tiharah
Ford
Entrenched Camp
Boondree
Goorsecan
Kote
Sand
Dry Nullah
JHOW JUNGLE
Two large Guns
D. Light Comp. and No. 6 sent out to watch Ford and to prevent a gun being taken by Enemy
Here also Capt. Holmes crossed
Here the greater Portion of the Enemy retreated.
References.
aa First British Line
bb Advance to Attack
cc Position after breaking Enemy's Lines
dd Sikh Cav.y threatening the left Flank.
ee. Col. Careton turning the left Flank of the enemy
ff Charge of 3rd Troops of 3rd L. Cy. under Maj. Hughes
gg - - the Right Wing of 16th Lancers
hh. - - - 3rd Squadron
mm - - - 4th -
nn. - - - 3rd Cavalry
oo. Advance of 2 Troops of H.A. with 53rd Foot and 30th N.I.
pp. Left Wing after driving the Enemy from Boondree.
qq. Village of Aliwal, stormed by 31st Foot and Nuss.ee Bat.n
rr. Batteries (not occupied.)
ss. Final Position of the British Cav.y after the Battle.
tt. Last Guns fired by enemy, taken by Ghoorkas.
vv. Line of retreat of part of Sikh Infantry
A. Last Guns fired by Sikhs
B. 50th encampment on the night of Jan. 28th.
C. 50th Camp on the 29th

The cavalry advanced in front with two troops of horse artillery, in the intervals of brigades, the infantry following in contiguous columns of brigades at deploying interval.

In this order the troops moved forward towards the enemy, who appeared to be directly opposite our centre, on a ridge about six miles off.

As the line approached the enemy, the cavalry were moved by brigades to the right and left, and took position in rear of both flanks, displaying the infantry columns.

The infantry made a brief halt under a heavy fire from the enemy about 10 a.m., after which the 1st Brigade, under Sir H. Smith himself, assisted by Brigadier Godby's Brigade, made a rapid advance, and captured the village of Aliwal and 2 guns; and Brigadier-General Cureton followed this up by a dashing charge of cavalry, which drove back the enemy's cavalry; after which Brigadier Godby's Brigade changed front, taking the enemy's line of entrenchments and camp in reverse.

Meantime the 50th Regiment, which was on the right of the 2nd Brigade, occupying the centre of the line (with heavy guns on each flank of the brigade) deployed and moved to their right, till they found themselves exactly opposite the centre of the enemy's position, which they at once advanced against.

The enemy's batteries being on a curve with the flanks thrown back, the 2nd Brigade in the centre were nearer and more exposed to the enemy's artillery than the other brigades. They were therefore twice ordered to lie down, to permit the brigades on each flank to advance in line with them.* In this position the regiment was exposed to a very heavy artillery fire, as well as to that of the matchlock men; but they advanced rapidly, and when within musket shot they fired a volley, which caused the retreat of the enemy opposed to them, the artillery alone remaining fast. The regiment then charged and took the guns opposite them, while the 16th Lancers on our left made gallant charges into the enemy's squares. These were soon in full retreat.

The "Queen's Own" Regiment continued the pursuit until they came near the bed of the river (see plan), No. 6 and the light companies were sent to the bank, to prevent a gun from being taken away.

Sir H. Smith says in his despatch dated—

"Field of Battle, Aliwal, 30th January, 1846.

"I occasionally observed Brigadier Wheeler's Brigade, charging and carrying guns and everything before it, again connecting his line and moving on, in a manner which ably displayed the coolness of the

* Lieutenant Bellars' diary gives a rather different account, as follows:—

"The artillery advanced and opened their fire. We then advanced up to them and lay down. This was done three times."

brigadier, and the gallantry of his irresistible brigade —H.M. 50th Foot, 48th Native Infantry, and the Sirmoor Battalion—although the loss was, I regret to say, severe in the 50th."

The enemy suffered very severely in this engagement, besides losing all their camp equipment, and all their numerous artillery except two guns, which alone escaped capture.*

The loss of the regiment in the above action consisted of—

Killed.

Lieutenant J. Grimes and 8 rank and file.

Wounded.

Captain Knowles, dangerously, and right leg amputated.
Captain L. Wilton, slightly.
Lieutenant Frampton, dangerously, left arm amputated.
Lieutenant R. B. Bellars, slightly.
Lieutenant W. P. Elgee, slightly.
Lieutenant A. White, severely.
Lieutenant W. Du Vernet, severely.
Lieutenant J. Purcell, severely.

* One of these guns must have been the one opposite to No. 6 and the light company of the 50th, concerning which Lieutenant Bellars says:

"No. 6 and the light company were sent to the bank to prevent a gun from being taken away from the opposite side. The Sikhs tried several times to take it, crawling on their hands and knees, but a volley always drove them away. The men were partially undressed to go over by the ford to bring it over, but Sir Harry would not allow them. They were relieved by a company of Sepoys who were left to prevent its being taken away, and in the morning it was gone."

Ensign W. R. Farmer, severely.
4 sergeants and 55 rank and file.

Loodiana being now safe Sir H. Smith rejoined the main army, leaving the 50th Regiment to watch the fords of that place. On the 3rd February, they encamped about a mile from their former cantonment at Loodiana, which had been left in charge of a small force of invalids quite inadequate to protect it; and it was found that some of the men's barracks, and the thatched houses of the officers, were burnt. All were plundered, and all the mess property of the regiment, to the value of nearly £3,000, was destroyed.

The same evening an express arrived ordering the 50th to rejoin Sir H. Smith at Sidham. They accordingly set out at 3 a.m. the following morning, and marching over the field of battle, rejoined the rest of the division that evening after a most fatiguing march of twenty-six miles.

The 31st and 53rd Regiments sent their bands out to escort the regiment in, and an extra glass of grog was served out to the troops, in honour of the regiment rejoining their comrades.

The whole force marched on the 5th February and joined the army under Sir Hugh Gough, in front of Sobraon, on the 7th.

Sir Hugh Gough rode out to meet the division, which formed line to receive him, when he complimented the 50th Regiment highly on their conduct at Aliwal.

It was generally anticipated that on the arrival of the heavy guns, which were daily expected, an attack would be made on the enemy's entrenched camp at Sobraon, which had now assumed a very formidable appearance.

On the evening of the 9th, an order was given for the whole force to be under arms at 3 a.m. the following morning without beat of drum.

Shortly after that hour, on the 10th, Sir H. Smith's Division moved silently through jungle across the country, and took up a position before daylight on the extreme right of the army, opposite the enemy's left, and with the right of the division resting on the river Sutlej.

In this position arms were piled.

The 31st Regiment and the 47th Native Infantry composed the 1st Brigade. The 2nd Brigade, consisting of the 50th, the 42nd Native Infantry, and the Mysore Battalion were in support.

SOBRAON.

About 6 a.m. our heavy guns opened on the enemy's lines, and after firing a few rounds took up a nearer position, from which a spirited cannonade was kept up, answered by the fire of the whole Sikh artillery.

About 11 o'clock the troops in the centre could be seen advancing against the enemy's centre, aided by shell and rockets thrown into their works. But so

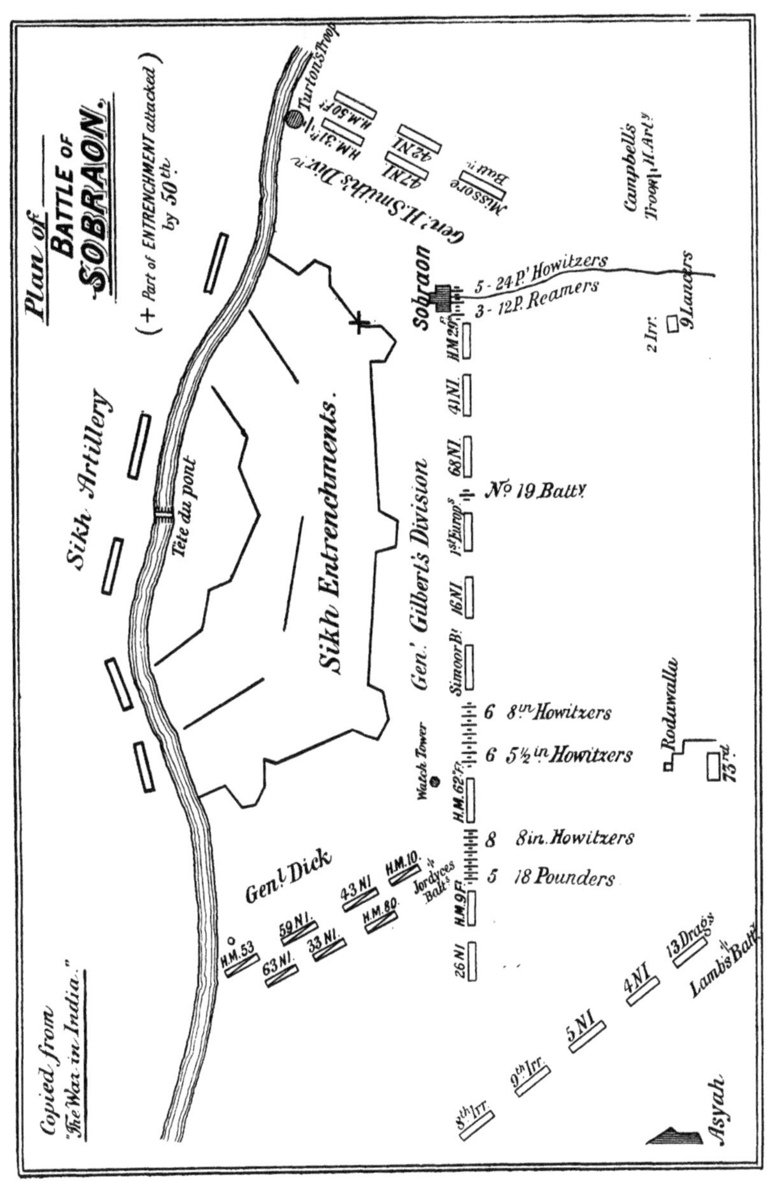
Plan of
BATTLE OF
SOBRAON.
(+ Part of ENTRENCHMENT attacked by 50th.)
Sikh Artillery
Tête du pont
Sikh Entrenchments.
Turton's Troop
H.M. 50 Ft.
H.M. 31 Ft.
42 N.I.
42 N.I.
Missore Batt.n
Gen.l H. Smith's Div.n
Campbell's Troop H. Art.y
Sobraon
5 - 24 P.r Howitzers
3 - 12 P. Reamers
2 Irr.
9 Lancers
H.M. 29
41 N.I.
68 N.I.
No 19 Batty
1st Europ.s
Gen.l Gilbert's Division
16 N.I.
Simoor B.n
6 8in Howitzers
6 5½ in Howitzers
Watch Tower
Rodawalla
73rd
H.M. 62 Ft.
8 8in. Howitzers
5 18 Pounders
Gen.l Dick
H.M. 10
Jordyce's Batt.s
43 N.I.
H.M. 80
59 N.I.
33 N.I.
H.M. 9 Ft.
H.M. 53
63 N.I.
26 N.I.
13 Drag.s
Lamb's Batt.y
4 N.I.
5 N.I.
9th Irr.
8th Irr.
Asyah
Copied from "The War in India."

severe was the enemy's fire that they were unable to effect an entrance. A simultaneous attack, however, which had been made by the Left Division, under General Sir Robert Dick, succeeded in forcing the enemy's line.

At this moment an aide-de-camp galloped up, and ordered Sir H. Smith to attack with his division.

The first line at once took ground to the left in column of sections, wheeling into line, and then advancing until opposite the position proposed to be attacked.

The 2nd Brigade followed in support of the above 100 yards in rear.

Both brigades soon got under a very heavy fire, and the 2nd Brigade were ordered to take ground to their left till clear of the left flank of the leading brigade, after which they steadily advanced under a most galling fire of shell, grape, and musketry. After passing a nullah, they were exposed to the fire of the whole Sikh batteries at musket range, but the Regiment pressed on gallantly, led by its commanding officers, Lieutenant-Colonels Ryan and Petit, who both fell desperately wounded within a few yards of one another.

The first brigade, after three most gallant attempts to force the enemy's entrenchments, were unable to do so, and retreated in confusion on the 50th Regiment, who formed fours deep with the steadiness of a parade movement,* and allowed the retreating troops to pass through their ranks. After this they re-formed line, and with a splendid cheer, rushed forward with the bayonet against the entrenchment, where they were gallantly met by the enemy, and a hand-to-hand struggle took place; but the 50th Regiment, proud of their former laurels, were not to be denied, and after a fierce and bloody struggle, they succeeded in forcing their way into the enemy's camp. In the meantime a large body of the enemy, driven back by the successful attack on the left, came

* I obtained this information personally from Major-General Thompson, who was present as a colour-sergeant. After his retirement I went to see him by appointment, to obtain all the information I could about this campaign. His first remark, which I give as nearly as I can remember in his own words, was—

"The thing that impressed me most in the whole campaign, was the steadiness with which the 50th at Sobraon, formed fours under a tremendous fire, to allow the 31st, who were retreating in disorder, to pass through their ranks, and then formed up as if on parade, and advanced, charging into the battery. You know how catching a panic is, and it struck me as a most trying ordeal."

This is reported without detriment to the 31st Regiment, who were mentioned in despatches for their gallantry on that occasion. They bore the heaviest part of the fire, which it was almost impossible to advance against.

Major-General Thompson is one of the few honourable instances of a private soldier, working his way up without interest, and should be an example to our young soldiers of the chances before them, with pluck, steadiness, and intelligence.

He enlisted in 1842, and was promoted to an ensigncy on the 27th February, 1852.

up behind, and turned the guns of the battery which the regiment had just taken upon them.

It was first thought that our own artillery had made a mistake, but directly it was ascertained where the fire came from, the 50th turned about and re-took the battery.

The enemy now retreated at all points, and the attacking divisions having concentrated, the bridge of boats was partially destroyed, and a heavy musketry fire opened on it. Our artillery raked it from each flank, and thousands of the enemy perished in their efforts to gain the opposite bank of the river.

When the 50th Regiment was formed up after the action, during the latter part of which it was commanded by Lieutenant Wiley, the senior subaltern,* it presented but a skeleton, nearly half the regiment that went into action in the morning having been put *hors de combat.*

* Lieutenant Wiley was afterwards promoted to an unattached company.

The returns were signed by Brevet-Major Long, who was not with the 50th Regiment during the action. He had been in command of a depôt, and arrived after it was over. The fact of Lieutenant Wiley being in command, is vouched for by a paper at the War Office, signed by Brigadier-General Penny, Major Low, and Lieutenant and Adjutant Bellars.

Bellars in his diary says—

"Sir H. Smith ordered me to collect the men outside the trenches, and the senior officer of the 50th was Lieutenant Wiley. Long took the command of the regiment, as the grog was being served out."

The following is a list of the killed and wounded :—

Killed.

Lieutenant C. R. Grimes.
Sergeant-Major Cantwell.
1 sergeant and 43 rank and file.

Wounded.

Major T. Ryan, K.H., Brevet Lt.-Col., dangerously.
P. T. Petit, Brevet Lt.-Col., dangerously.
Captain G. M. L. Tew, dangerously.
Captain J. B. Bonham, dangerously.
Captain H. Needham, dangerously.
Captain J. L. Wilton, severely.
Lieutenant H. W. Hough, severely.
Lieutenant J. G. Smith, severely.
Lieutenant C. A. Mouat, severely.
Lieutenant C. H. Tottenham, slightly.
Ensign C. H. Slessor, slightly.
8 sergeants, 1 drummer, and 177 rank and file.

Sergeant-Major Cantwell was killed just as the regiment had succeeded in forcing the enemy's entrenchment, and just as he had captured a colour from the enemy. One of the three Sikh colours, which adorned the mess of the regiment was captured here.

The Sikh army never recovered from this crushing defeat at Sobraon.

On the night of the 10th February, the 1st Division commenced their march to Lahore, where the conclusion of the war was announced, and the grenadiers of the 31st and 50th Regiments, were selected to furnish the guard of honour to the Maharajah Dhuleep Sing, on his visit to the Governor-General.

On the 20th the regiment was encamped at Meean Meer, three miles from Lahore.

Early in March the army was reviewed in the presence of the Maharajah, and the Governor-General of India, took the opportunity of presenting to the regiment, in a stirring address, Sir Charles Napier, now Governor of Scinde, who had so gallantly commanded them at Corunna.

Sir Charles Napier alluded in forcible terms, to the ties that existed between him and the regiment, and the imperishable glory connected therewith.

Some days afterwards he visited the camp of his old regiment; and on being shown the tattered remains of their old colours,* "he grasped them with enthusiasm, and expressed himself in glowing and characteristic language" (O.R.R.). It was a source of much regret to him, that he would be unable to present to the regiment the new colours now awaiting them at Loodiana, as he had to return to his seat of government.

The previous letter from Colonel Ryan to him

* These colours were presented on the 8th August, 1827, by H.M. King William IV. and Queen Adelaide, when Duke and Duchess of Clarence.

Some interesting information about Sir Charles Napier will be found in the Appendix.

on this subject, and his reply, are full of interest. (See pages 292-3.)

Lieutenant-Colonel Petit, who had commanded the regiment at the battles of Moodkee, Ferozeshah, and Sobraon, was appointed a Companion of the Bath.

The army was finally broken up on the 11th of March, and the regiment returned to Loodiana. Lieutenant-Colonel Ryan died at Kussowlie from the wounds received at Sobraon.

The Orderly Room Records say: "The deepest respect for his name must continue long after the present generation has passed away. His greatest ambition was to live, fight, conquer, and die at the head of his regiment."

He received his death wound commanding the 2nd Brigade, but while leading the 50th Regiment in one of its most glorious victories. He was appointed a Companion of the Bath, but did not live to hear of it.

About this time a subscription was raised to erect a monument in Canterbury Cathedral to those who fell in the campaign.

This monument consists of a female figure, emblematic of Fame, holding a laurel wreath in her right hand, in a reclining position, and resting on a large cannon, the breach of which only is visible. In her left hand she holds a standard, beyond which is partly visible "L," encircled by "Queen's Own." The figure of Fame was undraped in the original design, but the authorities of Canterbury Cathedral insisted on a draped figure, and it was altered accordingly.

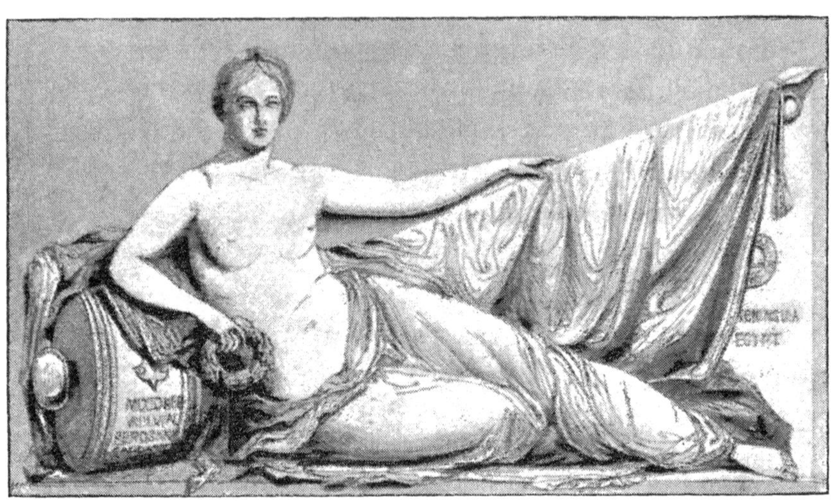

The inscription beneath the tablet is added, as it gives the names of all who fell in battle in this campaign.

Erected by the Officers of the 50th (Queen's Own) Regiment to the Memory of the Officers and Men who fell under the Colours of the Regiment during the Campaign on the Sutlej in 1845 and 1846, in the Actions of Moodkee, Ferozeshah, Aliwal, and Sobraon.

LIEUT.-COLONEL T. RYAN, C.B. and K.H.

LIEUTENANTS J. BROCKMAN, J. C. BISHOP, C. R. GRIMES, J. J. GRIMES, C. E. YOUNG.

ASSISTANT-SURGEON A. GRAYDON.

SERGEANT-MAJOR JOHN CANTWELL. COLOUR-SERGEANTS SAMUEL CHOVIL, JOHN WHYTE.

SERGEANTS MATHEW FENTON, ROBERT HENLEY, JAMES MCMACKING.

CORPORALS WILLIAM BECKET, GEORGE BRADSHAW, HENRY BROOKMAN, THOMAS CREAM, ROBERT FERGUSON, EDWARD JOHNSON, PATRICK KELLY, WILLIAM TAYLOR, SAMUEL TAYLOR.

PRIVATES

Astley, Robert.
Barry, William.
Baxter, John.
Black, John.
Bell, Thomas.
Beisley, Richard.
Boardman, Alexander.
Booth, William Henry.
Bowden, George.
Brooks, John.
Brennon, Peter.
Brindley, Joseph.
Brooks, George.
Brown, Josh.
Brown, Richard.
Buggins, Thomas.
Butterfield, Israel.
Butterly, James.
Cairns, William.
Callaghan, John.
Carroll, Dennis.
Clark, William.
Clifford, William.
Coley, Charles.
Colgan, John.
Connell, Dennis.
Connelly, Dennis.
Cook, James.
Coroner, William.
Cox, David.
Criswick, Morgan.
Cronnan, Bartholomew.
Cullen, Michael.
Cunningham, Thomas.
Daley, Eugene.
Daniel, Richard.
David, George.
Davis, James.
Davidson, John.
Dawson, Joshua.
Death, William.
Dempsey, William.
Densley, William.
Dennis, John.
Donlery, Michael.
Donaghan, James.
Donaghue, Phil.
Doolen, Daniel.
Dooley, Daniel.
Dooley, Michael.
Dowd, Michael.
Dunn, Francis.
Dunn, Richard.
Dyer, Thomas.
Eady, William.
Evatt, George.
Fanning, Peter.
Farney, George.
Ferris, John.
Feaney, Bryan.
Fitzpatrick, Daniel.
Flinn, James.
Garrett, William.
Garrett, William
Gibson, Richard.
Gillman, Richard.
Gleeson, Francis.
Goodwin, John.
Greenfield, George.
Hall, Anthony.
Halford, Benjamin.
Hamer, Aaron.
Hancock, George.
Harlow, Benjamin.
Harris, Fred.
Harris, James.
Hart, Henry.
Hayes, John.
Hayward, Edward.
Healey, Matthew.
Hogan, Daniel.
Holland, John.
Holwell, William.
Hughes, Ellis.
Husband, Robert.
Hutchin, John.
Hunt, John.
Ingold, John.
Jackson, Thomas.
Jackson, William.
James, John.
Jeffery, Joseph.
Johnson, Fred.
Johnston, William.
Jordan, Stephen.
Kelly, James.
Kelly, John.
Kelly, Dan.
Kelly, Martin.
Kenealy, William.
Keys, James.
Kelly, Philip.
Lawler, Peter.
Lawson, Thomas.
Leary, James.
Lucas, Benjamin.
Lynch, Michael.
Mahoney, Daniel.
Marsh, Richard.
Martin, William.
Mathew, Felix.
Mason, John.
McCable, John.
McCullam, James.
McGilvery, Peter.
McKean, Thomas.
McKenna, Patrick.
McNight, William.
Minney, William.
Mitchel, William.
Moran, James.
Mullins, William.
Murphy, Edmund.
Murphy, John.
North, Thomas.
Nugent, Michael.
O'Shaugnessy, Timothy.
Palfrey, James.
Pearson, Francis.
Peel, John.
Peg, Matthew.
Perry, Thomas.
Pike, Edward.
Pike, Thomas,
Poole, John.
Quigley, James.
Rasbottom, John.
Rees, John.
Reilly, Thomas.
Rice, James.
Richardson, John.
Richerbey, William.
Riddlestorff, Edward.
Roberts, George.
Roberts, William.
Ronalds, Richard.
Rutherford, Robert.
Shea, William.
Sheridan, J.
Skeats, John.
Smith, George.
Smith, John.
Smith, William.
Spanton, Robert.
Speak, Thomas.
Steele, Samuel.
Stevens, James.
Stevens, Thomas.
Stephens, George.
Sullivan, Dennis.
Thorpe, William.
Tomkins, Charles.
Travis, James.
Turner, George.
Wade, Francis.
Wade, Roger.
Wallace, Robert.
Walker, William.
Walsh, Edward.
Watts, Charles.
Wheaton, John.
Whitaker, J.
Williams, Edward.
Wilson, William.
Wiltshire, Peter.
Winn, Patrick.
Wright, William.

Monument in Canterbury Cathedral.

The following order, issued by Colonel Anderson on resuming the command of the regiment after this campaign, in which he was wounded, deserves to be preserved:

"They died as brave men like to die, and their memory and their names must, for the future, form a proud and conspicuous part of the records and history of this regiment."

A terrific storm swept over Loodiana on the 20th of May, and raged with such fury, that nearly the whole range of infantry barracks, ten in number, in which the 50th Regiment was quartered, were levelled to the ground, and a large number of the brave men who had come scathless through the war, lay mangled among the ruins.

The following station order of the 21st May, from the officer commanding the garrison, and the extract from the "Delhi Gazette" of the following day, gives full particulars of this disastrous incident:—

"It is with deep regret, the Commanding Officer announces to the troops at the station the calamitous disaster, accompanied with the fearful loss of life, of upwards of 80 persons killed and 135 wounded, which occurred yesterday evening to H.M. 50th Foot.

"During the height of the storm, the whole of the barracks of the regiment, including the hospital, were suddenly blown down, burying men, women, and children in the ruins.

"The loss of so many valuable lives so soon after

the arduous campaign, from which the corps returned crowned with victory and distinction, is doubly distressing, and the Commanding Officer feels sure that every man and officer will join with him, in deeply sympathising with the gallant regiment in its severe calamity."

From "Delhi Gazette," Loodiana, May 22nd, 1846:—

"This station has just been the scene of one of the most heartrending catastrophes ever witnessed. On the evening of the 20th a dreadful tuffaun, succeeded by heavy rain, razed the entire European barracks occupied by the 50th Regiment to the ground, sending upwards of 80 souls into eternity, besides inflicting most serious injuries on nearly 200 of the survivors. The scene which ensued baffles all description. The officers were in instant attendance, and exerted themselves to the utmost in extricating the unfortunate occupiers of the devastated buildings. The groans of the dying and the cries of the wounded were most harrowing. Husbands, wives, and mothers in agonised suspense awaiting the removal of the ruins, which was to reveal the mutilated remains of those most dear to them, all contributed a *tout ensemble* of the most heartrending woe, which will never be forgotten by those who witnessed it."

The evening and the greater part of the following days, were occupied in carrying away the dead; and at noon of the 21st, upwards of 80 bodies were ranged in the dead house.

On the evening of the 21st, they were buried without shroud or coffin, in three common graves near the fatal spot; where a monument has since been erected, commemorating the names of the victims.

At the beginning of December, orders were received for the 50th Regiment, to be held in readiness to embark for England, (before its proper turn on the roster); and it was authorised to allow up to 300 men to volunteer to remain with other corps. Of this 220 took advantage.

The remainder of the regiment commenced their march for Fort William, *viâ* Delhi, on the 18th December, and reached the former station on the 11th of April, 1847, having received *en route* nearly 100 volunteers from the 39th Regiment. An order was received from the Horse Guards, dated 30th June, 1847, authorising the action of Moodkee, Ferozeshah, Aliwal, and Sobraon to be borne on the colours of the regiment.

Colonel Anderson, who had been severely wounded at Punniar, when in the command of the 2nd Brigade of the 1st Division, rejoined the regiment from sick leave on the 3rd of February, 1847, and resumed command.

The 50th left Cawnpore on the 7th of February, and arrived at Benares on the 23rd of the same month, where the regiment embarked in country boats on the 25th for Calcutta; and, after passing through the Sunderbunds, arrived in safety off Fort William on the 11th of April. They landed on the following morning, and after marching round the captured Sikh guns (then parked in square beyond the Fort), the trophies of so much blood and glory, and which cost the regiment so many valuable lives, the gallant 50th passed proudly into Fort William, and there rested from its labours and fatigues, until the beginning of February, 1848.

Lord Gough issued the following General Order before the regiment embarked:—

"*General Order.*—Head-quarters, Simla,
22nd February, 1848.

"1. In directing the 21st and 50th Regiments to be struck off the strength of the Indian Establishment, consequent upon their embarkation at Calcutta for England, the Right Hon. the Commander-in-Chief in India takes the opportunity of congratulating these regiments, upon the satisfactory termination of their foreign service, and at the same time of recording the high opinion he entertains of their conduct, whilst under his command.

"The 21st Fusiliers, &c., &c.

"The 50th also commenced its tour of foreign service in New South Wales, and arrived in India in May, 1841. This regiment has been, as heretofore, most fortunate in being actively employed; highly distinguished in the wars of its country in

Egypt, and throughout the Peninsula campaign, the 'Queen's Own' has added to the numerous trophies already borne on its colours, the following glorious victories as records of its Indian service: 'Punniar,' 'Moodkee,' 'Ferozeshah,' 'Aliwal,' and 'Sobraon.'

"In the late eventful campaign the 50th bore a very prominent part; engaged in every battle that was fought, and conspicuous for its bravery amidst so many gallant regiments, nobly did it sustain on the banks of the Sutlej, the reputation it gained at Corunna, under its old commander, Sir Charles Napier. Every expression of Lord Gough's praise is justly due to this noble regiment, and that every happiness and success may attend all its members is his Lordship's most heartfelt desire.

"By order of the Right Hon. the Commander-in-Chief,

C. R. CURETON, Colonel,

Adjutant-General H.M. Forces."

The head-quarters embarked for England in "The Queen" on the 14th of February; and the rest of the regiment in the ships "Sutlej" and "Marlborough" * on the 4th and 15th of that month respectively.

The total strength of force embarked was 20 officers, 26 sergeants, 10 drummers, and 370 privates. The remaining officers were allowed to go home independently.

* The 6th is the date given in the Orderly Room Records, but the 4th is the date given in Sir H. Smith's despatch.

The "Sutlej," under the command of Brevet-Major Long, having on board also 1 captain, 2 subalterns, 1 assistant-surgeon, 7 sergeants, 3 drummers, and 73 privates, met with such serious damage on her homeward passage, that she was obliged to put into the Cape of Good Hope, and the detachment was landed while jury masts were being stepped.

Lieutenant-General Sir Harry Smith, Bart., who had commanded the division to which the 50th belonged in India, was at the time Governor and Commander-in-Chief of that colony; and previous to their re-embarkation on the 8th of May he issued the following general order:—

"H.E. the Commander-in-Chief has great pleasure in again expressing his marked approbation of the conduct of these men, his gallant comrades in India; for, to their bravery, fortitude, and exertions are to be attributed in a great degree the salvation of the ship, in a fearful hurricane off the Cape, carrying overboard the three masts in one crash.

"There are ties of friendship which war creates and cements, which nothing can sever; and Sir Harry Smith having again had brought to his notice the gallantry of his old comrades under appalling circumstances, derives as much pleasure in this record of his opinion, as he well knows the officers and soldiers of this detachment of the 50th (Queen's Own) will receive in its perusal."

This detachment landed at Gravesend on the

11th of July, the other detachment having previously landed there on the 6th of June; and a four-company depôt was formed at Walmer.

The Regiment moved to Dover in July; and on the 30th of November a letter was received from Lieutenant-General Viscount Hardinge, late Governor-General of India, stating that he would visit the "Queen's Own" to present to them a gold cup from Prince Waldemar of Prussia, who had served as a volunteer with that regiment in the Sutlej campaign, and who regretted having to leave England before he could present it in person. (See Appendix.)

In April, 1850, Major-General Lord FitzClarence presented to the regiment a handsome plate, to be attached to the Queen's colour, containing the names of those who fell under the colours in this campaign, and a crystal containing a piece of the silk of the colour; the staff on which it was fixed being the one presented by their Majesties King William IV. and Queen Adelaide.

CHAPTER IX.

CRIMEA.

The custody of the holy places in Palestine, had caused dissension between the Greek and Latin Churches. The Czar of Russia espoused the cause of the former, of which he was the head, while the Emperor of the French took the part of the latter. The Czar sent Prince Menschikoff to Constantinople, as ambassador to demand special powers, and backed up this demand by sending two army corps to the frontiers of the Danubian Principalities. The Sultan only gave in, to find himself confronted with still greater demands, which he could not grant; and on the 2nd of July, 1853, the Russian troops crossed the river Pruth, and entered the Principalities. The Great Powers attempted to mediate in vain, and on the 23rd of October a state of war broke out between the two countries. At first the Turks had some success—at Kalafat, and in the valiant defences of Silistria and Shumla. But it being now apparent, that the real object of Russia was the dismemberment of Turkey, France and England interfered; and on the 27th of February, 1854, they demanded the evacuation of the Danubian Principalities by the 30th of April, and this not being complied with, war was declared on that date.

An order was received from the Assistant Quarter-master-General, Dublin, on the 11th of February, 1854, stating that it was probable that the 50th Regiment would be immediately ordered on foreign service, and asking for returns showing the number of men required to complete to 850 privates. On the 18th of that month a further order was received directing the 50th Regiment to he held in readiness to embark on the 24th, to form part of an expeditionary force to be sent to the Mediterranean, to aid the Turks against the Russians.

The head-quarters and 6 companies, numbering 22 officers, 34 staff-sergeants and sergeants, 31 corporals, 10 drummers, and 612 privates, embarked on the steamship "Cambria," at Kingstown, Dublin, under the command of Major R. Waddy, on the 24th of February. They were disembarked at Malta on the 7th of March; and the "Cambria" returned to Kingstown. At the end of March, she embarked the remaining two companies of the regiment, numbering 9 officers, 11 sergeants, 9 corporals, 4 drummers, and 199 privates, which arrived at Malta on the 8th of April. Disembarking the detachments of other regiments she had brought out, and embarking two companies which had remained at Malta with the head-quarters of the 50th, she proceeded the next day to Gallipoli, where the troops landed on the 15th.

Meantime two companies, already under Major Wilton, had disembarked at Gallipoli on the 11th of April, and two under Major Möller on the 12th.*

The Regiment moved to an encampment near the village of Bulahar, on the 19th of April, being brigaded with the 2nd battalion Rifle Brigade, and the 93rd Highlanders. Here they were employed in throwing up lines of entrenchments, between the Gulf of Saros and the Sea of Marmora, to be used in case of a reverse, in the same manner as that in which the lines of Torres Vedras were used in the Peninsular War.

On the 6th of May the Rifle Brigade and 93rd Highlanders were moved to Scutari; and the 50th was moved to camp Chifleck, about two miles and a half from Gallipoli, where it was brigaded with the 1st Royals and the 38th Regiment, under the command of Brigadier-General Sir John Campbell, Bart., forming the 1st Brigade of the 3rd Division under Major-General Sir R. England, K.C.B.

On the 20th of June the muskets of the regiment were replaced by the Minie rifle.

On the 22nd of that month the "Queen's Own" embarked on board the "Cambria" and the transport

* The total strength of the whole regiment was 31 officers, 45 sergeants, 14 drummers, and 851 corporals and privates.

"Harkaway," the latter in tow; and, landing the sick and some women at Scutari, disembarked the remainder at Varna on the 24th. Here the regiment was encamped, first to the right of the Upper Shumla road, and afterwards on the 22nd of July near a muddy lake about half a mile north-west of Varna; and in these encampments the 50th, which had hitherto been very healthy, suffered severely from that fearful scourge Asiatic cholera, to which 24 men fell victims—but no officers.

On the 1st August they moved to some high ground near the sea, about half a mile north-west of Varna, and the spread of cholera was much checked.

The invasion of the Crimea having been determined on, the regiment was employed at this time in making gabions and fascines.

They embarked for the Crimea on the 2nd of September* on board four sailing ships in tow of the steamer "Sans Pareil." The whole expedition

* The subjoined is the strength carried by the various ships, the four first named being towed by the "Sans Pareil":

	Officers.	Sergeants.	Drummers.	Privates.	Total.
"Arabia"	5	8	6	110	129
"War Cloud"	6	5	1	107	119
"Tyronne"	3	6	1	104	114
"Shooting Star"	3	5	1	105	114
"Sans Pareil"	6	16	3	219	244
					720

left Baltchik Bay in fine weather on the 7th of September, all sailing vessels being towed by steamers; and on the 15th of September, the regiment was landed in boats about twenty miles south of Eupatoria, forming part of the army of invasion. No opposition was offered. They landed with three days' provisions and no baggage, except what was carried on their backs. The brigade was marched a short distance inland, halted in column, arms were piled, and every man was directed to lay down in rear of his arms. During the night it rained so heavily and continuously that all hands, having no great coats,* were speedily wet through. The next three days were occupied with the landing of stores, &c., and on the 19th, before daylight, the whole army was on the march. The three days' provisions issued on landing were exhausted, and no rations were issued before starting. The day was intensely hot, and there was no water except one little stream, which, by the time the 50th Regiment reached it, was almost black by the march through it of men and horses that had preceded them, but such as it was the troops were most grateful for it. Four halts took place during the day, after each of which the troops were marched over much ground, to take up a defensive position.

During the latter part of the day artillery was

* The great coats were landed next day.

heard in front, and the march was somewhat retarded. It was quite dark when, tired and hungry, the 50th arrived at their camping ground, on the banks of the little river Bulganac. There it was learnt, that a reconnaissance early in the afternoon, had shown the enemy to be in force on the other side of the river, where an affair of outposts had taken place.*

Late in the evening, after the rations had been drawn and the camp fires lighted, and when the men were hungrily looking forward to food and rest, an order came for the regiment to alter their position; and they were marched some distance to take up the new ground allotted to them, which was understood to be in preparation for a battle expected on the following day.†

Between 9 and 10 o'clock on the morning of the

* This march was very trying from the great heat, and want of food and water; while a peculiar aromatic odour arose from the bruised herbs trampled under foot, which at last became quite oppressive, and men were heard to offer to exchange grog for water.

Two men died from cholera.

In contrast to the heat of the days, the nights were intensely cold, and there was no protection from the climate for officers or men beyond the great coats, &c., which they carried themselves.

† The following is the formation of troops given by Kinglake:

"The first brigades of the Light and Second Divisions, were formed up parallel to the river, while the first brigades of the First and Third Divisions, formed an oblique line receding from the left of the Light Division, following the line of the coast, which here inclined backward somewhat sharply."

20th, the army moved steadily into the great plain, at the far end of which in the blue haze of distance, appeared a line of hills. The Turkish troops were on the right near the coast; and, covered by the guns of the ship, then came the French; on their left, the 2nd British Division in the first line, supported by the 3rd Division in the second line, while on the extreme left the Light Division in the first line was supported by the First; part of the 4th Division being in reserve, in echelon, in rear of the left, the remainder being placed in charge of stores at Eupatoria.

As the army moved further into the plain, the hills in front could be seen more clearly; and during occasional glimpses, the sunshine could be seen flashing on the bayonets of the enemy at the summit. As it approached nearer still, gardens and vineyards could be discerned, stretching down to a little stream with precipitous banks, skirting the base of the hill, which rose steeply beyond.

Presently, the Russian artillery opened fire, and the 2nd Division in front, which had been in double column of companies, deployed into line. A little further on the 3rd Division, supporting them, halted. The 50th Regiment retained their double-column formation, and piled arms.

In front of them the whole battle unrolled itself. To their right, the Zouaves could be seen crowding up the hill, like a swarm of bees. From the hill to the

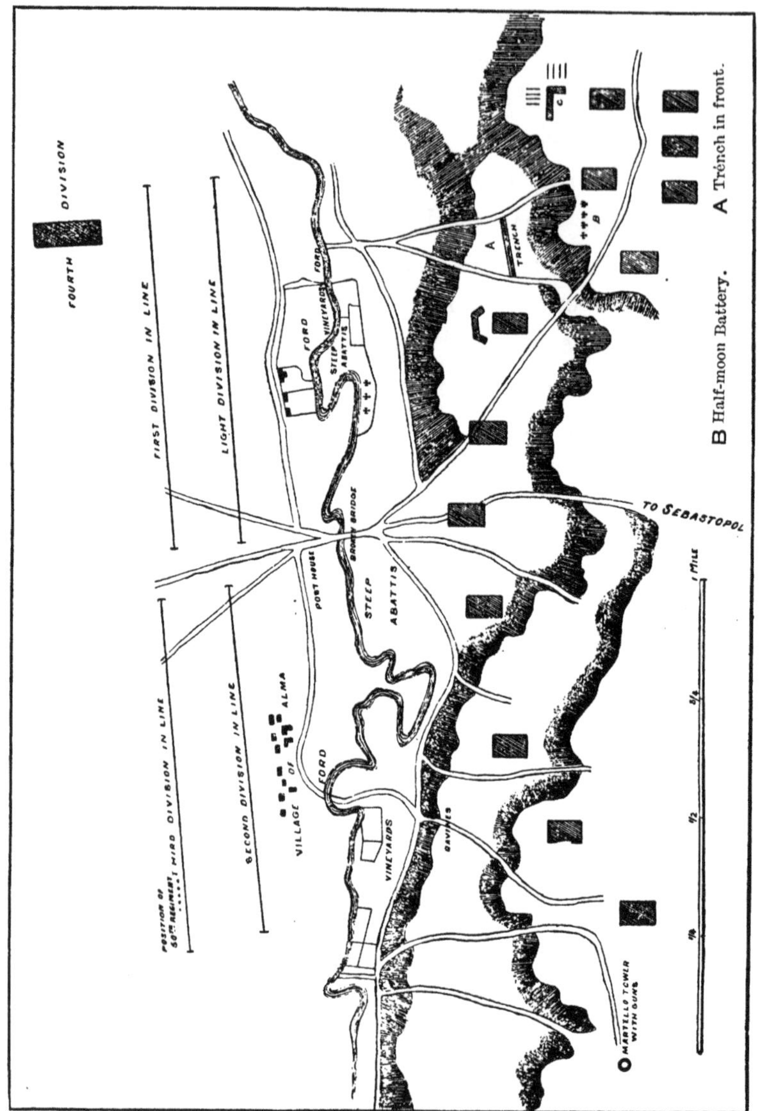

A Trench in front. B Half-moon Battery.

left front (afterwards known as the Half-moon Battery) came the continued roar of heavy artillery and musketry; while through the dense smoke of a burning village in front, occasional glimpses were caught of the steady advance of the thin red line. Occasionally a heavy shot flew over, or a spent one passed harmlessly through the ranks; and the horse of Major Watkin Wynne of the 23rd, who was killed, galloped, bloody and riderless, past the regiment.

At last the order came for the advance. Packs were taken off and left on the ground. Colours were unfurled and the regiment deployed into line, and moving steadily forward through the shallow stream, passed over the broken ground on the other side, encumbered by vineyards, where the number of the dead and dying told how fierce had been the struggle. Pressing up the steep hill, one harmless volley passed over the heads of the 50th, and they arrived at the top in time to find the Russian army in full retreat, and the whole country strewed with the packs, &c., which they had abandoned.

The British army bivouacked that night on the ground previously occupied by the Russians,* and the 50th Regiment was employed during the next three days in burying the dead, principally under

* They left in their bedding straw a legacy of vermin, which was unwittingly appropriated, and only got rid of when spare clothes could be obtained and the old ones boiled.

the Half-moon Battery, where the loss had been heaviest.*

On the 24th of September the army moved on to the Belbek, which appeared to be even a stronger position than the Alma, but was not defended.

On the 25th the regiment marched to Mackenzie's Farm, forming part of the celebrated flank march that led to the undefended south side of Sebastopol. Most of the way lay through a thick wood, with one road through it, which was ordered to be kept clear for the artillery and cavalry. The companies of the 50th, therefore, had to march through the wood itself, keeping the road in sight. In some places the wood was so thick that all formation was impossible, and all that could be done was to march independently, keeping the companies together.

Late at night, they emerged from the wood on to a steep decline, which led eventually to Mackenzie's Farm, where they halted for the night. This march had been a most distressing one, owing to the want of water; and several of the men were attacked with cholera.

The next day, after a short march, the regiment halted in a large vineyard full of ripe grapes, not

* Fatigue parties of the 50th Regiment were employed on the 21st in digging large pits for the burial of dead, under the Half-moon Battery.

In one place 7 officers of that gallant regiment, the Royal Welsh Fusiliers, lay dead, almost together, slightly in advance of a large number of their men who had fallen.

far from Balaclava; and the following morning they moved on to the plateau overlooking Sebastopol, from which, by the aid of a good glass, the town could be seen to be absolutely undefended.

On the 2nd of October the 3rd Division, to which the regiment belonged, took up its position on the left of the British army, where it remained during the whole of the siege, assisting to occupy the Greenhill Batteries (the left attack).

The ravine through which this division marched to the trenches, was generally known as the "Valley of the Shadow of Death," from the great number of cannon balls, which, just catching the high ground, rolled into the valley below.

Part of the approach was enfiladed, so that the reliefs for the trenches had to be marched off before daylight.

On the 2nd of October, a few tents were issued to the officers, and shortly afterwards the officers' bat ponies and the light baggage arrived at Balaclava; up to this time neither officers nor men had had a change of clothes, or any protection from the weather, except that which they carried themselves, and, though the days were very hot, the nights were intensely cold.

Lieutenant-Colonel Waddy was slightly wounded in the lip by a piece of shell, while serving in the trenches on the 13th. He was the first officer wounded in the trenches. Captain and Brevet-

Major Maxwell was wounded in the head on the 26th.*

Before dawn on Sunday morning, November 5th, the camp was aroused by the distant sound of heavy musketry, away to our right, towards the Inkermann valley. The day broke dull and foggy, with a misty rain, and before it was fully light the bugles had sounded the alarm and the regiment was under arms, mustering 17 officers, 26 sergeants, 10 drummers, 523 privates. By this time the firing on our right had considerably increased, and the artillery had joined in.

About 8 o'clock the 50th Regiment moved off, and, passing the Light Division camp, formed up near the Windmill.

Shortly afterwards Lieutenant-Colonel Waddy (in command) took up a position, with the left wing of the regiment on the Victoria ridge, while the right wing, under Lieutenant-Colonel Wilton (second in command), advanced in line towards the Inkermann ridge, and, doubling up, lay down behind the crest to the left front of the camp of the Second Division, a portion of which was still standing.

* Murchison, Maxwell, and the writer were standing together in the trenches, Maxwell, who was the tallest, being in the centre, when a round shot, just grazing the top of the parapet, passed close to Maxwell's head, knocking off his forage cap. He jocularly remarked, "That's a near shave," not knowing at first that some of the gravel from the parapet had been driven into his scalp. He had to be invalided to England.

The colour that accompanied the party in this advance, in charge of Lieutenant Antrobus, was covered with earth from a shot that ricocheted a yard or two in front.

From the crest behind which this portion of the regiment was lying down, the ground sloped gently to a level, beyond which rose Mount Inkermann, from the summit of which, about 1,200 yards off, the Russian guns kept up a rapid and continuous fire on our men, from which they suffered considerably, though lying down and out of sight. It was here that Lieutenant Dashwood was mortally wounded by a shell. He died that evening in camp.* This position was occupied by the right wing all through the action, and they remained on the ground as a guard all night, making their bivouac fires from the stocks of the Russian rifles.

Meanwhile the left wing, under Lieutenant-Colonel Waddy, which remained on the Victoria ridge, had

* During a lull in the fire of the Russian artillery, several officers of the right wing were walking about, which drew the fire of the enemy, and one of their shells burst just in front of the line, knocking over four or five men and wounding Lieutenant Dashwood mortally. One of his arms was carried away from the shoulder, and his water bottle and revolver were smashed, and driven into his side. An attempt was made to apply a tourniquet to stop the bleeding, but without success, and he died shortly afterwards. His body was brought into camp and buried there that evening, in the presence of all the officers, a round shot being place at each end of the grave.

one officer, Captain Frampton, wounded, and lost several men from artillery fire.*

At the close of the action, when the enemy's heavy artillery could be seen retreating towards the Karabel Faubourg of Sebastopol, by the West Sappers Road, Lieutenant-Colonel Waddy saw an opportunity, which he immediately seized, of trying to cut them off.†

With this view he descended into and crossed the Careenage ravine with No. 7 company, about 60 strong, under Lieutenant Murchison, and "making his way up the north-west angle of Mount Inkermann, he advanced through a covert of brushwood, and took up a position within easy distance of, and commanding the flank of the retreating column, and overlooking the head of the harbour. The movement was most opportune, for he reached the position just as the guns, wagons, and great teams of draught horses were jammed into one confused mass, and when the infantry, not expecting to be attacked so near home, were taking a route lower down."

Murchison's company at once opened fire upon the

* Captain Frampton had previously lost an arm at Aliwal, and was one of the two officers taken prisoner at the sortie from Sebastopol on the 21st of December.

† The description of this dashing attack is taken from Kinglake's "Crimea," and from the statements of an officer who accompanied No. 7 company. Kinglake adds, "The enemy, considering that his hampered train of artillery was helpless against infantry, and prevented by brushwood from detecting the paucity of his assailants, believed for a moment that the enormous prize was on the point of falling into their hands."

enemy; and though, owing to the very wet weather, only about half the rifles would go off, the confusion of the column was manifestly increased. Had a sufficiently strong force then been employed, the Russians would have been in danger of losing much of their heavy artillery.

"General Todleben at once recognising the danger sent a company in skirmishing order, supported by 4 battalions of the Bourbaki Regiment 3,000 strong and 4 guns" to attack Murchison's company. Against such overwhelming odds it was hopeless to contend, and Lieutenant-Colonel Waddy, seeing that the retreat of the company was in danger of being intercepted, ordered them to retire as quickly as possible. The men then retreated independently the best way they could into the ravine, the descent into which was very precipitous. An ineffectual fire of musketry and artillery was opened upon them, but they made good their retreat with the loss of one man taken prisoner.

The 50th was the only regiment of the Third Division engaged in the above battle.

A terrific storm swept through the camp on the 14th November, carrying away nearly every tent, and causing great destruction of shipping and stores in the harbour of Balaclava, the "Black Prince," which contained all the winter supplies of clothing, &c., going down among others.

On the 21st December, the trenches of the

left attack being held by the 50th Regiment, the Russians took advantage of the darkness of the night, to make a sortie in force. Moving silently, in dense masses, they overlapped our left of the advanced trench, which was held by Captain Frampton, Lieutenant Clarke, and 150 men. They first came in contact with the advanced sentries on the left of the third parallel; these at once gave the alarm by firing, and retired into the trench. The enemy then pressed boldly forward between the second and third parallels. The latter, being attacked both in front and rear by overwhelming masses, was taken; and both the officers and 9 men were taken prisoners, after many of the men had been killed and wounded, Lieutenant Clarke being among the latter; a few of the defenders making good their retreat to the second parallel, which was now furiously assailed. Here Major Moller was mortally wounded, dying that evening; and many of the 50th were killed and wounded. But the enemy were repulsed at all points.

By this time the heavy firing had alarmed the camp, and reinforcements having been at once sent down, the advanced trench was retaken, and the Russians retired, leaving behind them 14 killed and 4 wounded. The loss of the 50th Regiment consisted of 14 killed, 12 wounded, and 2 officers and 9 men taken prisoners.

General Sir Richard England called on Lieutenant-

Colonel Waddy the next day, to convey the thanks of Lord Raglan, for the successful defence of the trenches, and for the gallant conduct of officers and men of the regiment.

The regiment suffered severely during the winter months. During the first six months of the siege 340 men died of disease, and 8 officers were invalided.

Enfield rifles were issued on the 14th June, 1855.

The 50th Regiment with detachments of the 14th and 39th Regiments, were told off as the storming party of the Barrack Battery, at the attack on Sebastopol of the 18th June, but owing to the failure of the attacks on the Garden Battery and the Redan, which commanded both flanks, the attack was impracticable.

A portion of the regiment, which held the advanced trench on this occasion, suffered from the heavy fire from Sebastopol.

Sir Richard England having been invalided, Sir William Eyre assumed the command of the 3rd Division on the 2nd of August.

"The Queen's Own" was in reserve at the final attack on Sebastopol on the 8th September, which culminated in the fall of the south side.

From this date fatigue parties were constantly employed in the town, and frequently exposed to fire from the guns on the north side.

A tremendous explosion occurred in the French camp on the 15th of October; their magazine blew up, and a falling shell set fire to the British magazine also. The dense smoke hung like a pall in the sky, while hundreds of live shells were in the air at once, carrying death and destruction around. It was fortunate that the conflagration did not extend to the windmill, which was used as a powder magazine, or the casualties would have been still greater; as it was, at least 500 men were killed or wounded, and all the English right siege train, and most of the French, with guns, tumbrils, huts, and horses, were destroyed.

Toward the end of November the winter had fairly set in, with a good deal of rain and occasional frost and snow. At first men and officers were in tents, after a time three huts were erected, and early in December huts were built for all the men, and a large one to hold 16 officers was put up, for the accommodation of those who had not built for themselves.

The trenches were dismantled and the guns sent on board ship in readiness for all exigencies in the spring. Splendid roads were constructed, 10,000 to 12,000 men being employed on them as working parties every day, and abundance of stone being available.

The south side of Sebastopol was held by the French and English conjointly, the former taking that part which lies to the west of Woronzoff road, the latter the part to the east of it.

Arrangements had been made for blowing up the docks, and it was calculated that that magnificent work, a marvel of construction, which is reported to have cost twenty millions, and to have taken 40 years to build, could be destroyed in ten minutes. An enormous number of Russian guns, captured when the south side fell, now lay idle in their embrasures.

The health of the regiment was good, though a few of the men had suffered from exposure, and it remained good throughout the winter, a great contrast to the previous winter.

The strength of the 50th Regiment on the 1st January, 1856, was 28 officers, 31 serjeants, 9 drummers, and 511 privates.

They were inspected by General Sir William Eyre, G.C.B., on the 31st January, who was much pleased with their appearance and efficiency.

At the end of January the Russians took advantage of a dark night, to make a reconnaissance across the harbour in boats, but they were discovered and fired on by the sentries, and at once retreated, with the exception of one boat, which was discovered later, and also compelled to retire. This drew a very heavy fire from the north side; an eye witness says 135 shots were counted in five minutes, and I saw 6 or 7 shells in the air together.

On the 24th of February a grand review of British

infantry was held by General Codrington, appointed commander-in-chief when General Simpson, who succeeded Lord Raglan, went home. 20,000 bayonets in splendid order marched past the English commander, who was afterwards joined by Marshal Pelisier, commanding the French army.

A soldier was hung for murder at the end of February, detachments of 150 men from every regiment being detailed to witness the execution, which took place early in the morning in pouring rain.

An armistice for one month was concluded on the 1st of March, and on the 2nd April information was received that a peace had been signed in Paris on the 30th of March.

Preparations for evacuating the Crimea were at once made.

The embarkation of the British army commenced early in May, and on the second of that month, the 50th Regiment moved to the Marine Heights, on the south side of Balaclava Harbour, to assist in the embarkation of stores.

The head-quarters and right wing of the regiment embarked on board H.M.'s steamship "Royal George" on the 10th July, and the remainder embarked in H.M.'s steamship "Algiers" from Balaclava Harbour on the 12th of that month.

Previous to the departure of the latter vessel a guard of the 50th Regiment, under Captain Thompson, handed over Balaclava to the Russians, and the

English sentries were relieved by Russians; Captain Thompson being the last British soldier to leave the Crimea.

Besides the Crimean medal and clasps, the following officers, non-commissioned officers, and men of the regiment received distinctions:—

CROSS OF COMPANION OF THE BATH.

Lieutenant-Colonel R. Waddy.

CROSS OF KNIGHT OF THE LEGION OF HONOUR.

Lieut.-Colonel . . .	R. Waddy, C.B.
„ . . .	J. L. Wilton.
Major	H. I. Frampton.
Brevet-Major . . .	A. C. K. Lock.
Colour-Sergeant . .	Angus McPherson, No. 2823.
Private	Thomas Regan, No. 2933.

SARDINIAN MEDAL.

Lieut.-Colonel & Colonel .	R. Waddy, C.B.
Major Bvt.-Lieut.-Colonel .	H. E. Weare.
Major	E. G. Hibbert.
Lieutenant . . .	N. A. Clarke.
„ . . .	J. Lamb.
Private	Andrew O'Leary.
„	Thomas Regan.

MEDJIDIE.

Lieut.-Colonel & Colonel .	R. Waddy, C.B.
Brevet-Lieut.-Colonel .	H. E. Weare.
Major	E. G Hibbert.
Brevet-Major . . .	A. C. K. Lock.
„ . . .	D. W. Tupper.

Captain . . . . E. C. Antrobus.
" . . . . J. Thompson.
Lieutenant . . . M. A. Clarke.
" . . . J. W. Lee.

French Military War Medal.

1789 Sergeant-Major . . Robert Foley.
2783 Colour-Sergeant . . William Turner.
3253 Sergeant . . . . R. W. Newcombe.
3810 Private . . . . John Brennan.
3903 " . . . . William Cooney.
3606 " . . . . Michael Hannon.
3500 " . . . . Lawrence Ward.

Medal for Distinguished Conduct in the Field, with Annuity or Gratuity.

1289 Quartermaster-Sergeant . Thomas Clifford.
1624 Colour-Sergeant . . Joseph Duncalf.
1871 " . . George Kent.
3340 Corporal . . . . William Fahey.
2263 " . . . . John Golding.
2823 " . . . . Andrew O'Leary.
2070 " . . . . Richard Rogers.
1348 Private . . . . Edward Cade.
2736 " . . . . John Daniels.
3368 " . . . . Daniell Flynn.
3609 " . . . . Alexander Grant.
3057 " . . . . Jerem Moran.
1599 " . . . . Patrick O'Brien, 1st.
3042 " . . . . Patrick O'Brien, 2nd.
2629 " . . . . James Quinn, 1st.
2150 " . . . . John Wait.
3156 " . . . . John Walsh, 1st.

CHAPTER X.

THE NEW ZEALAND WAR.

THIS war commenced by a dispute about some land at Waitara in 1859, when some desultory operations took place, and a truce was made in 1861.

But it broke out again in 1863, when Sir George Grey, the governor, sent soldiers to occupy Tatarai-maka, which the Maories had seized, and in May of that year Lieutenant Tragett, Dr. Hope, and 8 rank and file fell into an ambuscade, and were barbarously murdered. The hostile natives then threatened Auckland and other European settlements; and General Cameron was recalled from Taranaki, with most of his force, for the protection of Auckland.

The most important operations, previous to the arrival of the 50th Regiment, were the dispersal of a strong column of the enemy, that moved against General Cameron on the Koheroa range, and the capture of the strong position of Meri-Meri, on the right bank of the Waikato River, at the end of October.

On the 15th of November of this year, the 50th Regiment landed at Auckland, under the command of Colonel Waddy, C.B., and marched next day to Otahuhu, new colours having been presented previous to landing by the colonel's wife.

The hostile natives, reinforced by the fugitives from Meri-Meri, having taken up a strong position at Rangariri, a native village on the right bank of the Waikato River, about 12 miles above Meri-Meri, General Cameron attacked them on the 20th November, assisted by two iron steamers, and, after repeated assaults on the enemy's works (owing to the troops being unprovided with scaling ladders), only resulting in a useless sacrifice of life, the place, on being surrounded, surrendered unconditionally the following day; and on the 8th of December Ngaruawahia was taken possession of unopposed. Here General Cameron was detained, waiting for supplies, till the 27th December.*

The 50th Regiment meantime occupied strong positions as follows: Head-quarters and 3 companies at Drury; 2 companies at Queen's Redoubt; and 1 company at each of the following stations: Shepherd's Bush, Martin's Farm, Williamson's Clearing, and Razorback.†

* Interchangeable rifles were issued to the 50th Regiment on the 2nd of December, 1863.

† About this time Colonel Waddy, C.B., was appointed colonel on the staff. He was afterwards acting brigadier, until the 8th April, 1866.

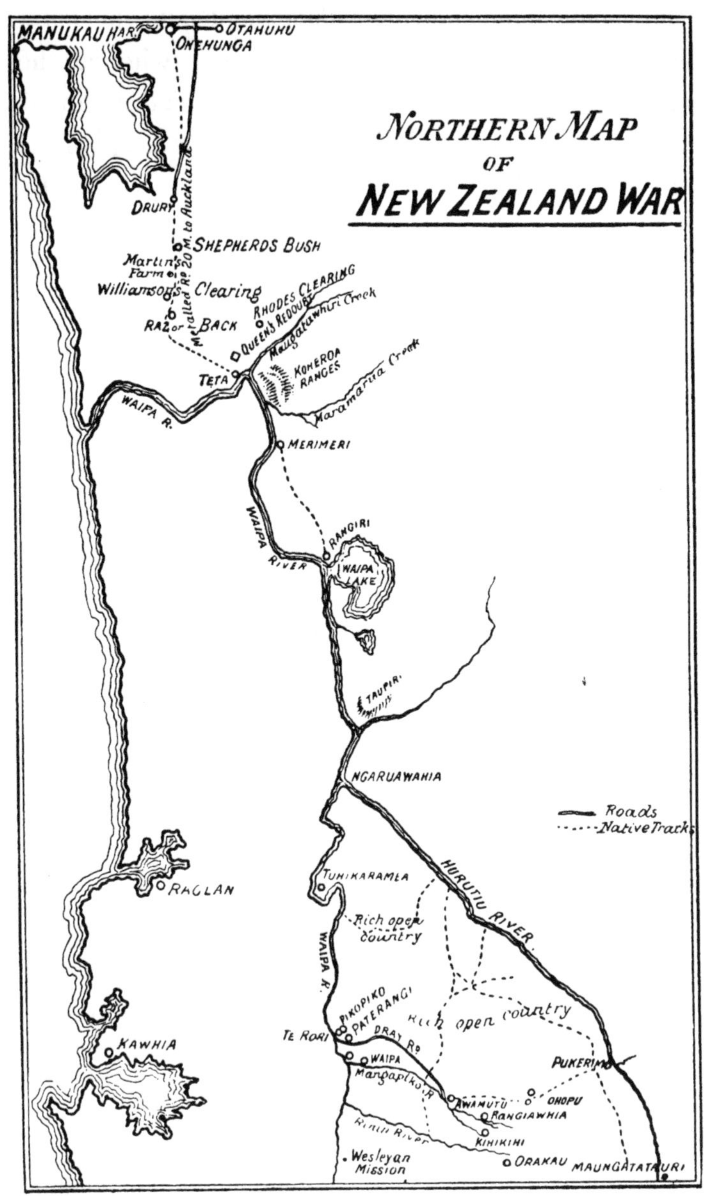
NORTHERN MAP
OF
NEW ZEALAND WAR
MANUKAU HAR.
OTAHUHU
ONEHUNGA
DRURY
Metalled Rd. 20 M. to Auckland
SHEPHERDS BUSH
Martin's Farm
Williamson's Clearing
RHODES CLEARING
QUEEN'S REDOUBT
Mangatawhiri Creek
RAZ or BACK
KOHEROA RANGES
TETA
Maramarua Creek
WAIPA R.
MERIMERI
WAIPA RIVER
RANGIRI
WAIPA LAKE
TAUPIRI
NGARUAWAHIA
Roads
Native Tracks
HURUTIU RIVER
TUHIKARAMEA
RAGLAN
Rich open country
WAIPA R.
PIKOPIKO
PATERANGI
Rich open country
KAWHIA
TE RORI
DRAY RD
WAIPA
Mangapiko R
AWAMUTU
OHOPU
PUKERIMU
RANGIAWHIA
KIHIKIHI
Wesleyan Mission
ORAKAU
MAUNGATATAURI

The regiment received orders on the 20th December to concentrate at Otahuhu, from whence they marched to Onehunga, from which place 5 companies embarked on the 4th January, 1864, for Raglan, and were followed by the rest of the regiment on the 6th. This party landed at Raglan at 6 p.m., and found that the first part of the regiment had gone into the interior, and was encamped about 10 miles off. The officer in command (Captain C. R. Johnson) being unable to obtain a guide, or any information as to the road, beyond a vague description in broken English from some Maori women, marched off his party in the direction which they indicated. After an hour's marching the road became a series of tracks, branching in various directions. Taking the one which seemed the most used, the party soon became benighted, lost the track, and bivouacked in the fern for the night. At daylight the next morning, they found themselves about five miles off the road leading to the camp.

The regiment then marched to the head of the Waitatura Valley; and the men were employed in making a road over the ranges, so as to form a junction with the force under General Cameron, then advancing along the right bank of the Waipa river.

On the 26th January the regiment crossed over by this road to Tuhikaramea, leaving Lieutenant-Colonel Hamley, with 440 men, on the left bank; the remainder of the regiment crossed the Waipa river, and joined General Cameron's force. On the following day General Cameron advanced along the right bank, Lieutenant-Colonel Hamley's party co-operating on the left bank ; and on the 28th this latter party also crossed the river, and the whole force took up a strong position on a hill at Te Rori, on the left flank of the Pateranghi Pah, and about 200 yards from the river, which was navigable for small steamers.

On the 11th February, about 3 p.m., a party of the 50th Regiment, under the command of Captain Doran, accompanied by some men of the 40th Regiment, were proceeding to bathe in the Mangapiko creek (a tributary of the Waipa), when they were fired on by the enemy, concealed in fern on the opposite side. The party returned the fire, and reinforcements were sent up till 200 men were engaged under Sir H. Havelock. After much hot firing the troops succeeded in crossing the Mangapiko by a bridge formed of a single plank, though the banks of the river were forty to fifty feet high, and densely wooded ; and a series of hand-to-hand encounters took place in the thick bush, resulting in the Maories being completely routed.

The 50th Regiment had 2 rank and file killed.

At noon on the 20th February General Cameron marched to Rangiawhia, which he seized at daylight on the 21st, surprising the enemy in their beds. The 50th Regiment, which he left at Te Rori, joined him the next day at Awamutu, to which place he moved after the capture of Rangiawhia. On the 22nd the enemy were observed moving from Pateranghi; and Colonel Waddy, C.B., who now held a separate command as colonel on the Staff, advancing on that place with 120 men from his Corps of Observation, found it was evacuated. Simultaneously Sir H. Havelock found Piko-Piko had been abandoned. The same day it was found that a large force of natives, who had retreated from Pateranghi, were entrenching themselves between Awamutu and Rangiawhia; and General Cameron at once sent a force against them. The Maori position extended for about 400 yards along the crest of a ridge at right angles to the road to Rangiawhia, which it crossed, and the site of an old pah afforded advantageous cover.

On the approach of two companies of the 70th Regiment, sent on as advanced piquets, the enemy's skirmishers, thrown forward a mile from their entrenchments, opened fire at 300 yards from behind a hedge, but were speedily dislodged.

While two guns and skirmishers were engaged with the enemy, the 50th Regiment, under Colonel Weare, forming the assaulting column, were lying down in the road waiting for orders.

They were supported by the 65th and 70th Regiments. On the order for the assault being given, the 50th Regiment dashed forward under a heavy fire, but until close to the trench and rifle-pits, the enemy's position could only be approached by a four-deep formation along a narrow road, hemmed in on either side by high ferns and impenetrable manuka scrub, which opened out into clear ground about a hundred yards in front of the position. Lieutenant White (50th Regiment) was sent forward with 20 men as a

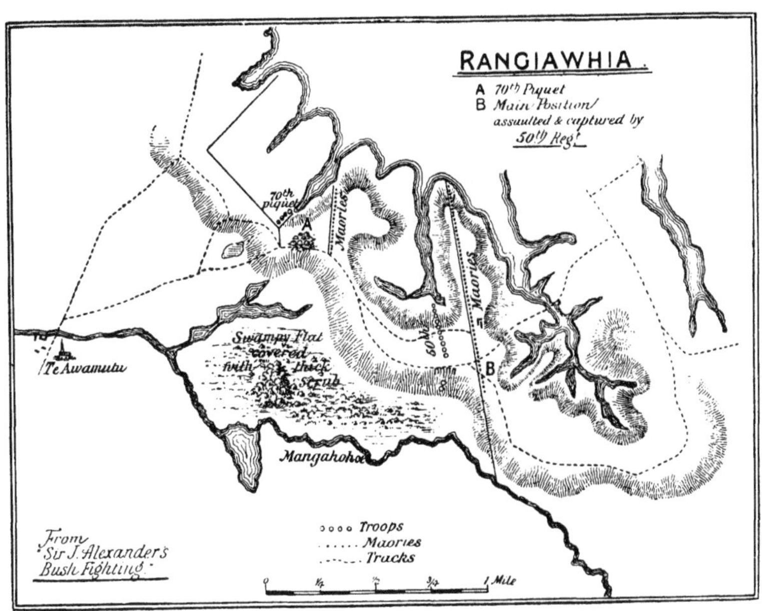

false attack to draw the fire of the enemy, being quickly followed by No. 1 Company, under Captain Johnson, and No. 10, under Captain Thompson, which took ground to the left and right respectively, so as to turn the flank of the enemy, the remainder of the regiment attacking their front. The enemy were found posted behind a post-and-rail fence, and a bank with ditches on either side. Being unable to withstand the impetuosity of the bayonet charge of the 50th Regiment, they fired a parting volley when within twenty yards, and, evacuating the rifle-pits, broke and fled towards Mangatautari Mountain, leaving 40 killed and 4 wounded on the field.

The casualties of the regiment were 1 sergeant killed, Ensign Doveton, 1 sergeant, and 9 privates wounded.

General Cameron complimented the regiment on the field, for the splendid manner in which the position had been taken, and the following is extracted from his despatch on the subject to Sir George Grey, the Governor:

"Headquarters, Te Awamutu, 23 Feb., 1864.

"The 50th were exposed to a heavy fire in advancing towards the position, which they carried with great gallantry.

"I beg to enclose a copy of Colonel Weare's report describing the attack.

"I cannot praise too highly the admirable conduct of all the troops, Regular and Colonial, during the fatiguing night march of the 20th, and the operations of the two following days; but particularly of the mounted Royal Artillery under Lieutenant Rail, and of the 50th under Colonel Weare.

"I have, &c. &c.,

"(Signed) S. A. Cameron, Lieut-Gen.

"H.E. Sir G. Grey."

"Extract from Colonel Weare's despatch—

"'The nature of the ground and formation left little for the commanding officer to do but to place the men, in the first instance, and then leave officers commanding companies to fight their men, and I am proud to say that officers and men nobly did their duty under very trying circumstances, and while exposed to a fire that must have caused a very large increase to the list of casualties, had it not been for the dense dust raised by the men doubling, which partially concealed them.

"'I beg to bring to the notice of the Lieutenant-General commanding the forces, the names of Captains Johnson and Thompson and Lieutenant White, 50th Regiment.

"'I much regret to say that Ensign Doveton, 50th Regiment, fell dangerously wounded by the side of Captain Thompson, while gallantly performing his duty.'"

On the 22nd of March General Cameron left Te Awamutu with some Royal Artillery, the 50th and 70th Regiments, the Forest Rangers, and 50 men of the Naval Brigade, and encamped at Pukerimo, on the left bank of the Horatui river, about seven miles from Maungatautari, a strong position taken up by the force of Maories which had been driven from Rangiawhia on the 22nd February. On the 23rd, this force advanced to within 300 yards of Maungatautari, and found the enemy posted in two very strong positions, consisting of earthworks constructed on spurs of the Pukekera Range, which completely blocked the road. These earthworks were well flanked and palisaded. The force having reconnoitred the position returned to camp.

On the 14th General Cameron, having been informed of a native path practicable for infantry, which led to the village in rear of the position, advanced the Forest Rangers by that route. The enemy's scouts having warned them of this advance, they retired into the Taupo country, evacuating both positions, which were occupied by the 50th Regiment, who remained in charge of them.*

The field force was now broken up, and the 50th Regiment remained in camp in the Maungatautari district until the 15th August, when the headquarters proceeded to Otahuhu, leaving a detach-

* On the 21st March Kaitaki fell, and on the 31st an unsuccessful attempt was made to take Orakua by assault; it was again assaulted on the 2nd of April, when the Maories abandoned the position, and escaped into the bush, leaving 120 killed and 30 wounded behind them.

Pukehinehana having been occupied in considerable force after the retreat of the Maories from Maungatautari, a force was moved against it on the 27th of April, and on the 28th some troops passed through the shallow part of a mud flat to the rear of the position. On the 29th the guns and mortars opened fire on the pah, and at 4 p.m. on that date an assault was ordered; after a fierce fight the assailants were repulsed with heavy loss, especially of officers; but during the night the place was abandoned by the enemy. The 50th Regiment was not engaged in any of the above operations.

ment at Pukerimu, which rejoined the regiment in October.

In December the Regiment embarked at Onehunga for Wanganui.

On the 24th of January a field force, consisting of the 2nd battalion 18th Regiment, the 50th Regiment,

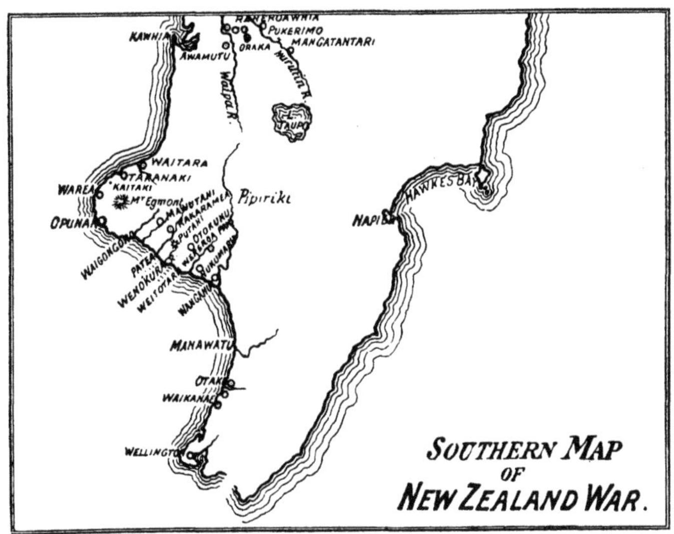

and detachments of the 57th, under Brigadier-General Waddy, C.B., marched into the interior of the country and encamped at Nukumaru, about two miles from the rebel position of Wereroa.

The right piquet held by the 18th Regiment was at once attacked by the enemy.

At 2 p.m. the next day the left piquet, held by Lieutenant Wilson and 80 men of the 50th Regiment, was attacked by a strong force of the enemy, which continually increased, and, after a stubborn resistance, the piquet was forced to retire; but being quickly reinforced, an engagement took place which lasted about two hours, the enemy being eventually driven back to the pah.

Eleven privates of the 50th Regiment were killed, and 2 officers (Lieutenant Wilson and Ensign Grant) and 18 privates were wounded.

February 4th, Brigadier-General Waddy, with 1,104 officers and men, was ordered to cross the Waitotara River, leaving Colonel Weare in command of the men left at Nukumaru.

February 15th, Wereroa appearing to be very strong, Brigadier-General Waddy was ordered to march to Patea, in the hope of drawing the enemy from their pah, and Colonel Weare took up Brigadier-General Waddy's camp at Waitotara.

March 9th, Colonel Weare's party rejoined the column at Patea.

March 13th, the column, consisting of 1,273 officers and men, moved up the right bank of the Patea River towards Kakaramea. After marching about 2½ miles the enemy was discovered skilfully posted on the hills commanding the right flank. They opened fire at a range of about 300 yards, and clung tenaciously to their very strong position.

The advanced guard, composed of the 57th Regiment under Major Butler, then changed direction to the right, and supported by a portion of the 68th Regiment under Lieutenant-Colonel Morant, and a company of the 50th Regiment under Captain C. R. Johnson, (the remainder of the regiment, under Major Loch, being in reserve), advanced upon the enemy and drove them from their position after a stout resistance. The original march on Kakaramea was then resumed, and the village taken by Captain Johnson's company, after a sharp skirmish, and afterwards occupied by a detachment of the regiment.

The enemy's loss in killed and wounded is variously estimated at 40 and 60 by different authorities, but even the latter seems a low estimate, as 43 bodies are reported to have been found.*

Captain Leach, 50th Regiment, who acted as Deputy Assistant Adjutant-General on this occasion, received the thanks of the general.

The following day the force moved northwards,

* My authority for stating that 43 bodies were found is the *Wanganni Chronicle.*

Major Barker, who was adjutant of the regiment at the time, writing to me on the subject, says:

"I think this number" (43 bodies) "is pretty accurate, as there were 5, one a woman, on the hill where they opened fire on us; 7 in the swamp behind; 2 killed by mounted men; 3 or 4 on the slope leading to the village; several in the fern between the village and the river, and for many days afterwards bodies were found floating down the river, and bodies were found also lying in various directions by patrols and wood-parties. One of the prisoners we took escaping from Botahi, told me there were 50 Maories who never came back after that fight."

meeting with no further opposition, when Manutahi was occupied and found to contain a large quantity of stores, including 200 tons of potatoes.

The regiment successively occupied Tangahoe on the 29th March, Waingongoro on the 3rd of April, and Waimate on the 9th of April.

The winter season coming on, General Cameron decided on advancing no further, and the regiment returned to Patea. The head-quarters and 400 men being formed into a moveable column, were encamped on the right bank of the river.

On the 2nd June, 1865, Colonel Weare's column marched northward from Patea, in order to form a junction with Colonel Warre's force marching southward from Taranaki.

This junction being satisfactorily effected, and the country found to be quiet, Colonel Weare's party returned, arriving at Patea on the 13th June.

A letter from the general officer commanding was received, expressing satisfaction with the manner in which this duty was performed, and with the report of the country supplied by Lieutenant and Adjutant Barker, 50th Regiment, who acted as D.A.Q.M.G.

Shortly afterwards, Sir George Grey, having set his heart on the capture of Wereroa, took on himself the direction of affairs. He determined to occupy the heights around the pah, and to shut in the defenders. Brigadier-General Waddy was posted with 400 regular troops about 900 yards from the position; but this party was only to be employed as a "moral support." On the 20th of July Lieut.-Colonel Trevor, 14th Regiment, and Captain Noblett, 18th Regiment, each with 100 men, pitched their tents about 1,300 yards in front of the pah. On the same date a force of 400 men of the Native Contingent moved north, in heavy rain, towards the Karaka Heights, leaving Colonel Trevor's and Captain Noblett's parties alone in front of the pah. They reached the native village Aciari in the evening, and surprised it, making prisoners of the inhabitants just before daylihgt next morning. From this village a narrow bush track led to the plateau in rear of Wereroa pah. At daybreak on the 21st rifle shots into Wereroa from the Karaka Heights awakened its defenders. About 9 p.m. the whole force marched through the village of Perikama, and climbed the steep heights leading to the pah. On the way up they met a native, who told them it was deserted; this they found to be the case, one old woman being the only occupant.

The prisoners taken during this operation were shipped off to Wellington, confined in a hulk about a mile and a half from the shore, and placed in charge of an officer and some men of the 50th Regiment.

On the 25th of August Sir Duncan Cameron, having applied to be relieved on account of ill-health, sailed for England, and was succeeded in the command by Sir Trevor Chute.

Major-General Sir Trevor Chute, who now commanded the troops in New Zealand, moved from Wanganui to Wereroa on the 30th of December, 1865; and on the 3rd of January, 1866, the troops moved through the Waitotara block to the strongly fortified pah of Otokuku, which the rebels considered impregnable. That afternoon the Native Contingent, under Major MacDonald, entered the pah by surprise and burned the wahries or native huts. The following morning, however, the pah was reported to be occupied in force, and General Chute made dispositions to attack it. After a march of about 2 miles through dense forest, and almost impracticable ravines, the troops gained the plateau on which this formidable defence stood; and it was gallantly captured by a charge of the 14th Regiment, driving the enemy down the almost scarped sides and rear. Information had already been sent to Colonel Weare, 50th Regiment, commanding at Patea, and Major Lock, 50th Regiment, commanding at Manavoupo, instructing them to patrol the country in the vicinity of their posts, with a view of intercepting and cutting off the enemy's retreat.

The force marched to the Wennokura river on the 6th of January, and encamped on some high land to the south, and in front of the rebel stronghold of Putahi. Four companies of the 50th Regiment were directed to advance on it, at the same time from the Patea side, next the river. Putahi is situated on a clearing, about half a mile in diameter, on the top of a hill rising abruptly from the river, to the height of about 500 feet, and covered to the crest with dense bush. The usual approach was on the side on which the main force was encamped, and was the one the rebels evidently anticipated we should attempt, they having erected stockades, and other impediments to assist in its defence. It was, therefore, decided to attack it in the rear, and General Chute, having succeeded in obtaining the services of a guide, marched his force at 3 a.m. on the 7th, crossed the river by a bridge constructed the evening before, and, passing over the plain, ascended a steep spur which brought them on a plateau to the left front of the pah. The march may be described as one of continued struggle through ravines and gullies, which could only be ascended and descended by the aid of the supplejacks, and then with great difficulty. The extreme distance was barely four miles, but the obstacles met with made the march last over four hours. When the force arrived at the clearing, the Native Contingent, who had led through the bush, formed to the left, and the Forest Rangers opened out, lying down in skirmishing order, to cover the formation of the remainder of the force, who, as they emerged one by one from the bush, were extended with supports—the 2nd Battalion 14th Regiment in the centre, 2nd Battalion 18th Regiment on the right, and the 50th Regiment on the left, the Native Contingent forming the reserve. This formation occupied more than an hour, under a desultory fire from the pah, from which it was distant about 400 yards, and when complete the order was given to advance. The rebels now opened a heavy fire, but the line advanced steadily and did not charge until within 50 yards, when, with a cheer and rush, they carried the position, the enemy retreating to the bush beyond, where the Native Contingent was ordered to pursue them.*

The whole of the troops behaved admirably, and although working on broken ground, conducted the attack as steadily as on an ordinary parade.

Colonel Weare, commanding the 50th Regiment, had previously been directed to send 100 men from the Patea to the right bank of the Wennokura river; this party succeeded in intercepting some of the rebels retreating by the inland road towards Kakaramea, and taking several prisoners.

The Regiment moved from the Patea district to Taranaki on the 22nd of February, and was stationed in the following redoubts: Opunaki, on the coast about half way between Patea and Taranaki; Warea, about thirty miles; and Stony River, about twenty-six miles from Taranaki, where the head-quarters were established.

On the 26th of August, the head-quarters and right

* The account of the capture of Putahi is taken from General Chute's despatch of January 8th, 1866.

wing received orders to proceed to Auckland, for embarkation to New South Wales. They arrived at Auckland on the 17th of September. On the 26th of that month one company embarked for Brisbane; and on the 3rd of October the head-quarters and four companies embarked for Sydney, which they reached on the 9th.

Five companies of the left wing, under Major and Lieutenant-Colonel F. G. Hamley, continued to hold the above-mentioned redoubts, but the war was virtually over.

On the 3rd of June, 1867, two more companies embarked for Sydney; and on the 11th of July the remaining three companies, under Lieutenant-Colonel Hamley, sailed for Adelaide, South Australia, which they reached on the 9th of August.*

Brevet-Colonel Richard Waddy, C.B., who had commanded the 50th Regiment throughout the Crimean war, was subsequently promoted to the rank of major-general on the 11th of November, 1868.†

A Horse Guards Order of the 1st July, 1870,

* Major and Brevet-Lieutenant-Colonel F. G. Hamley, commanding a detachment of the regiment at Adelaide, South Australia, was appointed Governor in Chief and Captain General of that Colony, on the 19th of February, 1868, consequent upon the death of Sir Dominick Daly, K.C.B., and administered the Government until February, 1869.

† General Sir Richard Waddy, K.C.B., died at Kingstown on the 9th July, 1881, and a monument was erected to his memory in Canterbury Cathedral by his brother officers.

authorizes the words "New Zealand" to be borne on the colours of the regiment, in commemoration of its services in that country during the years 1863, 1864, 1865, and 1866.

The regiment remained in Australia till the 24th of March, 1869, when the head-quarters embarked at Sydney, on board the "Himalaya," which had previously taken up the Brisbane detachment. They arrived at Adelaide on the 29th, where they embarked the detachment under Colonel Hamley. The regiment disembarked at Devonport on the 14th of June, 1869.*

Little that is of interest occurred during the home service of the Queen's Own in England, Ireland, and Scotland until the year 1881. In April of that year the 50th was one of the four regiments made up to a total strength of 1,040 of all ranks.

It would be a grave omission to close this history of the 50th Regiment, without a reference to the colours, which it has borne so honourably through many vicissitudes.

The earliest official information, that I can find of new colours having been received, occurs in the

* During its service in Australia the regiment was repeatedly reduced, and finally on the 19th September, 1868, its strength was :—

Service Companies	-	620 rank and file.
Depôt Companies -	-	80 ,, ,,
Total	-	700 ,,

Inspection Report of the 50th, drawn up by General Dilke in Dublin, on the 27th July, 1768, in which he notes, "Two colours, new in 1763;" * but standards must have been presented to the regiment before this—probably to the 52nd, before it became the 50th.

The 1763 colours were repaired in 1769 or 1770, and replaced by new ones (sent out to Gibraltar) on the 3rd April, 1786. A note in one of the inspection reports might lead to the belief, that standards were again given to the regiment in 1777, in which year they were reported as bad; but from subsequent reports I am of opinion that they were repaired in 1777 or 1778, as they are again reported as bad in 1781, 1784, and 1785, which would not have been the case if new ones had been received in 1777, while repairs in that year would just have kept them in fair condition till 1781, which is implied by no mention being made of them in the intervening returns. Colours were again received in 1804, on receipt of which the old colours were burnt, with military honours, at the head of the regiment. This is the first colour noted in the Orderly Room Records.

The illustration on next page is taken from a picture at the Heralds' College, to which the following note

* The date of these new colours corresponds with the return of the regiment from Germany after the Seven Years' War (March, 1763), during which campaign the previous colours would have been worn out.

from Major Haven,* commanding the regiment, is attached:

"Cork 1807. Colours made some time ago."
"The Sphinx and Egypt added lately."

This drawing must, therefore, have been taken from the 1786 colour, to which the "Sphinx" and "Egypt" had been added. The colours of 1804 must have borne these devices from the beginning, as they were sanctioned by official letter of July 6th, 1802. Nevertheless, with these devices it becomes a true representation of the 1804 regimental colour. These colours were borne with honour in the campaigns of Copenhagen, Vimiero, Corunna (where Ensigns Moor and Stewart were shot under them), in the Walcheren Expedition, and through the Peninsula.

On new colours being presented in 1815, the old ones, that had seen such splendid service, were burnt with military honours at the head of the regiment. Thus following the precedent that had been set with the 1786 colour, which was borne in Corsica and through the 1801 Egyptian campaign. New colours were again supplied in 1815;† these and the 1863

* Major Haven became senior Major of the 50th Regiment on the 18th June, 1801.

† An article in the January number of the "Journal of the Royal United Service Institution" says, "that on the 50th receiving new colours in 1827 from Queen Adelaide, then Duchess of Clarence, the old colours which had been carried in the Peninsula War were cremated, and the ashes preserved in a box forming part of the mess plate, and on the box was

standards were the only ones that never were in action.

This illustration represents the regimental colours (of the pair) given to the regiment, by His Royal Highness the Duke of Clarence, on the 8th August, 1827. It is peculiarly interesting, as it was in consequence of this, that the 50th Regiment became "the Duke of Clarence's Own" in the following September, and was authorised to wear his device, W with a crown over it, at the three corners of the regimental colour (authority at the Heralds' College). It does not appear, however, that this device ever was actually borne.

All the above colours consisted of a broad red St. George's cross on a black ground (with blue war scrolls), to match the black facings of the regiment; and though the 50th became the Queen's Own in 1831, and assumed blue facings as a royal regiment, the colours were not changed; and the pair presented by the Duke of Clarence were borne with great distinction, through the campaigns in India. These were the colours that were shown to Sir Charles Napier in 1845 when "he grasped

engraved the names of those who fell under the colours in action." This must refer to the 1804 colour, not to the 1815 one, which preceded the one presented by the Duke of Clarence in 1827.

I wish some confirmation could be obtained of this. I have a faint recollection of hearing a rumour of this kind, when I first joined the regiment, but it was only a rumour, and there is no foundation for the story of the mess plate.

them with enthusiasm and expressed himself with characteristic and glowing language." (O.R.R.)

In 1845 new colours were sent out to India, and Colonel Ryan, commanding the regiment, endeavoured to get Sir Charles Napier to present them in the following letter:—

"SIR,—

"On the 8th of August, 1827, his Majesty King William the Fourth, presented in person and in the presence of the Queen Dowager of England, the colours of the 50th (Queen's Own) Regiment, which are now borne in its ranks.

"The sacred banners are now, by the lapse of time, faded and mouldering. They have been present in one battle since their presentation, that battle is now enrolled, by the Sovereign's command (in addition to others), on those standards. You have, Sir, served in the 50th Regiment, and commanded it in one of its fiercest fights, and you obtained, at its head and in its front, early fame and renown.

"New colours have now reached the regiment at this station. They are yet unfurled.

"I now write to you, Sir, in the name of that regiment, which I have the honour temporarily to command, and I also, as the last remaining representative and soldier of the old 50th, invite you to present (when a favourable opportunity may occur) the new colours.

"Banners received from the hands of a soldier, whose name is associated with the Peninsula 50th in bright and glorious days, cannot fail to increase in the young 50th, that *esprit de corps*, which has long been characteristic of the 50th Regiment, and which cannot fail to animate, and cheer to glorious results, when they are first unfurled on the battle-field.

"I have, &c., &c.,

"J. RYAN, Lieut.-Colonel,

"Major Commanding 50th Regt.

"LOODIANA, 8th August, 1845."

To this Sir Charles Napier replied:

"KURRACHEE, August 22nd, 1845.

"SIR,

"I received your letter, written to me in the name of the 50th Regiment, and asking me to present to it its new standard.

"Tell the regiment, I pray you, that every feeling expressed by me in dedicating my book on military law to the corps is still vivid in my heart. I gloried in its ancient laurels. I glory in those it more recently gained at Punniar. Proud shall I be, to present its new colours to this celebrated regiment, which victory has always attended; for its valour ever seized with ardour the opportunities offered by its fortunes.

"If fate throws us together before the 16th January, 1846, great will be my satisfaction; if not, I must resign, with regret, the honour you intended for me; but I shall remember with pride, that on that day thirty-seven years ago, I commanded the 50th in battle under the great and immortal Moore, whose dying eyes were fixed on the advancing colours of our regiment.

"I have, &c.,

"NAPIER, Major-General."

Unfortunately Sir Charles Napier had no opportunity of presenting these colours, having to return to his government of Scinde, and the 50th Regiment continued to bear the old ones, embarking under them for England in 1848. It was on the occasion of the embarkation of the head-quarters from Fort William, that the general officer commanding, reported of these colours: "Hardly a rag remaining. They have nobly done their duty."

The 1845 colours were taken into use for the first time at Walmer Barracks, Deal, on the 18th July, 1848. These were the first colours with a blue ground ever borne by the regiment. They were carried through the Crimea, and continued in use until new ones were received on board ship, which were presented by the wife of Colonel Waddy, commanding, on the 5th November, 1863, previous to the regiment landing for service in the New Zealand War.

TABLE OF COLOURS AND ACTIONS.

Year.	Where borne.	Year.	Where borne.
About 1756	7 Years' War.	1815	—
1763	—	1827	Sutlej.
1786	Corsica. Egypt, 1801.	1845	Crimea.
		1863	New Zealand.
1804	Copenhagen.		
	Vimiero.		
	Corunna.		
	Walcheren.		
	Peninsula.		

On the 1st of July, 1881, territorial titles came into force—and the 50th Regiment became (The Queen's Own) Royal West Kent. The 97th Regiment, which at one time also bore the title of "The Queen's Own", becoming the 2nd battalion.

In the following February Lieutenant-Colonel Fyler, commanding the regiment, presented a cup formed from the staves of the old Crimean colours set in gold with a crystal stem containing a portion of the colour itself. This last record of the Old Colours fittingly closes the history of a regiment whose gallant deeds are "footprints in the sands of time."

APPENDIX I.

SHORT SKETCH OF THE LIFE OF SIR CHARLES NAPIER, G.C.B.

CHARLES NAPIER, the eldest son of Colonel the Honourable George Napier and Lady Sarah his wife, was born in 1782, and educated in Ireland.

When he was twelve years old, he was appointed to an ensigncy in the 33rd Regiment.

He was appointed to the staff at Limerick in 1799; he was afterwards for some years on half-pay of the 4th Irish Regiment, became captain of the Royal Staff Corps, obtained a brevet majority on the 29th May, 1806, and was posted to the second battalion of the 50th Regiment, on the 6th November of that year.

He writes from Ashford, Kent, in March, 1807:

"Our men have got ophthalmia, and are dying fast; also from inflammation caused by the coldness of the weather, and bad barracks. There is no raging fever—cold alone is the cause—yet the men die three or four a day."*

Again in May he writes:

"The soldiers have got pneumonia at Hythe, and are dying as fast as we folks at Ashford. Only think of a surgeon taking in one day one hundred and sixty ounces of blood, and the man is recovering! They say bleeding to death is the best way of recovering them!" *

* Both these quotations are from Butler's "Life of Sir C. Napier," a book that should be read by every 50th officer.

The above extracts, taken by permission of Messrs. Macmillan and Co. from Sir W. Butler's "Life of Sir C. Napier," show that at the age of twenty-five, Charles Napier took a deep interest in the soldiers' welfare.

Major Charles Napier was transferred in 1818 from the second battalion of the 50th to the first battalion, then on active service in Spain, in order to relieve Major Hill, wounded at Vimiero, and shortly afterwards, on the promotion of Colonel Walker, he obtained temporary command of the regiment, which he held through all the difficulties and trials of the retreat, and the battle of Corunna, which are fully dealt with in the chapter on that campaign.

The following extract from Captain Patterson's work, shows how Major Napier obtained the love of his soldiers, on the trying march from Monte Santo to Salamanca:

"The wearied soldiers toiled with difficulty along, under the most tempestuous weather, severely felt in those Alpine regions, where the cold was so excessive, as to require the hardest bodily exercise to withstand it. In order to keep the men alive, the band and drums were frequently put into requisition, and our commander, Major Napier, occasionally ordered some well-known national quick-step, when in a moment, as if by magic, the tired and jaded, stepped out once more with fresh spirit.

"The young recruits and drummers felt the hardship most, and often on the journey has Major Napier given his charger to one of them, or to any poor fellow who could not well get on, while with a musket, or sometimes a brace, on his shoulders, he walked before the regiment."

The account of the capture of Major Napier at Corunna, given in his own words in Sir W. Butler's "History of Sir C. Napier," is so full of interest that I give it verbatim, by permission of Messrs. Macmillan and Co.:

"I said to the four soldiers (Irish privates of the 50th and 42nd), 'Follow me, and we will cut through them.' Then, with a shout, I rushed forward. The Frenchmen had halted, but now ran on to us, and just as my spring was made the wounded leg failed and I felt a stab in the back; it gave no pain, but felt cold, and threw me on my face. Turning to rise, I saw the man who had stabbed me making a second thrust, whereupon, letting go my sabre, I caught his bayonet by the socket, turned the thrust, and, raising myself by the exertion, grasped his firelock with both hands, thus in mortal struggle regaining my feet. His companions had now come up, and I heard the dying cries of the four men with me, who were all instantly bayoneted. We had been attacked from behind by men not before seen, as we stood with our backs to a doorway, out of which must have rushed several men, for we were all stabbed in an instant, before the two parties coming up the road reached us. They did so, however, just as my struggle with the man who had wounded me was begun. That was a contest for life, and being the strongest I forced him between myself and his comrades, who appeared to be the men whose lives I had saved, when they pretended to be dead on our advance through the village. They struck me with their muskets, clubbed and bruised me much, whereupon, seeing no help near and being overpowered by numbers and in great pain from my wounded leg, I called out 'Je me rend,' remembering the expression correctly from an old story of a fat officer, whose name being James, called out 'Jemmy round.' Finding they had no disposition to spare me, I kept hold of the musket, vigourously defending myself with the body of the little Italian who had first wounded me; but I soon grew faint, or rather tired. At that moment a tall, dark man came up, seized the end of the musket with his left hand, whirled his brass-hilted sabre round, and struck me a powerful blow on the head, which was bare, for my cocked-hat had fallen off. Expecting the blow would finish me, I had stooped my head in hopes it might fall on my back, or at least on the thickest part of the head, and not on the left temple. So far I succeeded, for it fell exactly on the top, cutting me to the bone, but not through it. Fire sparkled from my eyes. I fell on my knees blinded, but not quite losing my senses, and holding still to the musket. Recovering in a moment, I saw a florid, handsome, young French drummer holding the arm of the dark Italian, who was in the act of repeating the blow Quarter was then given; but they tore my pantaloons in tearing my watch and purse from my pocket, and a little locket of hair which hung round my neck But while this went on two of them were wounded, and the drummer, Guibert, ordered the dark man who had sabred me to take me to the rear. When we began to move, I resting on him because hardly able to walk, I saw him look back over his shoulder to see if Guibert was gone, and so did I, for his rascally face made me suspect him. Guibert's back was towards us; he was walking off, and the Italian again drew his sword, which he had before sheathed. I called out to the drummer, 'This rascal is going to kill me; brave Frenchmen don't kill prisoners.' Guibert ran back, swore furiously at the Italian, shoved him away, almost down, and putting his arms round my waist supported me himself. Thus this generous Frenchman saved me twice, for the Italian was bent upon slaying.

"We had not proceeded far up the lane when we met a soldier of the 50th walking at a rapid pace. He instantly halted, recovered his arms and cocked his piece, looking fiercely at us to make out how it was. My recollection is that he levelled at Guibert, and that I threw up his musket, calling out, 'For God's sake don't fire; I am a prisoner, badly wounded, and can't help you; surrender.' 'For why would I surrender?' he cried aloud, with the deepest of Irish brogues. 'Because there are at least twenty men upon you.' 'Well, if I must surrender—there,' said he, dashing down his firelock across their legs and making them jump, 'there's my firelock for yez.' Then coming close up he threw his arm round me, and giving Guibert a push that sent him and one or two more reeling against a wall, he shouted out, 'Stand back, ye bloody spalpeens, I'll carry him myself; bad luck to the whole of yez.' My expectation was to see them fall upon him, but John Hennessey was a strong and fierce man, and he now looked bigger than he was, for he stood upon higher ground. Apparently they thought him an awkward fellow to deal with He seemed willing to go with me, and they let him have his own way."

Major Napier was released on parole not to serve till exchanged, which did not permit him to rejoin his regiment, then in the south of England, till January, 1810, but home service had little attraction for him, and in May of that year he volunteered to the staff of General Crawford, commanding the Light Division in the Peninsula. He had two horses shot under him at the attack on Crawford's position in front of Almeida, and was afterwards severely wounded on the staff of Lord Wellington at Busaco, the day after the 50th landed at Lisbon (see page 127). He was invalided at that place throughout the winter, during which the army occupied the lines of Torres Vedras.

Early in March when the campaign reopened in the spring, though still suffering from his severe wound, he went off to join his regiment; riding 90 miles in 24 hours (Sir W. Butler), which he must have reached about the time of the investment of Almeida, with the subsequent action of Fuentes d'Onor.

In June, 1811, Major Napier was promoted to the command of the 102nd Regiment.

He left the 50th Regiment on the banks of the Tagus, and they saw him no more until they met at Lahore, in India, 35 years later (see page 236). How much had happened in that interval. The 50th Regiment were fresh from their recent victories in the Punjaub, and Sir Charles Napier was the renowned conqueror of Scinde.

For a succinct account of his later services I cannot do better than refer to the speech of Lieutenant-Colonel H. de B. Sidley, commanding the 50th Regiment, when Sir Charles dined at their mess in Preston on the 19th June, 1852.

The letter of Sir Charles Napier accepting Lieut.-Colonel Sidley's invitation to dinner is so characteristic that I give it in full:

"May 31st, 1852.

"MY DEAR SIDLEY,

"The Officers of the Fiftieth do me great honor, and I shall have much pleasure in dining once more at my dear old mess. Few are now alive that were at it in my time, but there are some. It was somewhere about the time of the Deluge! but whether before or after I don't quite remember. I think after, because as we were 'The Devil's Royals' Noah would not have let a man of us into the ark.

"Well, deluvian or antideluvian, I will with God's blessing dine with you all on the day after the Preston dinner, *i.e.* the 19th June. I must be back by the 20th; will you settle this with Mr. Cooper, who has been so kind as to ask me to his house. I wish you joy of the command of a Regiment which, if antideluvian, may have run away before the flood, but certainly never did since! Well, give my best respects to the glorious old 50th. I hope my abominable liver will not prevent my seeing you on the 19th, for I love the Regiment as if I had never left it.

"I have my old belt and breast-plate still, and I will wear it when I go to you, as it is to me a memorial of old and glorious days.

"Yours truly,
"C. NAPIER."

I give below an extract from the newspaper report of the above. Sir C. Napier's account of the death of Sir John Moore will be found full of interest:—

"On Saturday evening last, Sir Charles Napier's stay in the neighbourhood permiting it, a banquet was given at the barracks by the officers of the 50th Regiment to 'the Conqueror of Scinde,' to commemorate his visit to Preston, and in order that they might have once more at their mess-table, the hero of a thousand fights. The room was decorated with irrefragable evidences of the bravery of the 50th, and of their having been engaged in hot action. Behind the chairman's seat were the fragments of two colours, which had often waved over the scene of conflict and of victory, during the campaign in which the 50th were engaged in India. These flags certainly bore unmistakeable proofs of having seen hard service, the remnant of them consisting of nothing more than the poles and a few tatters—very rags. On the other side were three other colours, trophies of war, originally the property of the Sikhs, from whom they were taken by the 50th. On the table was a silver cup, presented to the 50th by Waldemar, Prince of Prussia, who was travelling in India during the campaign, and joined the 50th Regiment, to which he became much attached. He is since dead.

"Sir Charles Napier was escorted by a guard of honour, consisting of a hundred men, each wearing a medal, and commanded by Captain Frampton, of the light company, and who lost an arm at the battle of Alliwall. The other officers over the guard of honour were Burns and Thomson.

"Lieutenant-Colonel Sidley rose and said, 'I rise with very much pleasure, but with much diffidence, to propose the toast of the evening—diffidence, from a knowledge that the pleasing office has fallen into feeble hands, and pleasure at feeling how highly honoured the officers of the 50th and myself are, at having once more amongst us, at his own mess-table, our distinguished guest (Cheers). It is impossible by anything I can say, to add to the well-earned fame of the conqueror of Scinde. He is, however, dear to us all, as the major who commanded the 50th Regiment throughout the Peninsula campaign, terminating with the battle of Corunna (applause) under the distinguished Sir John Moore, whose dying eyes were fixed upon the advancing victorious banners of the 50th, and whose last words relating to our distinguished guest were, "Well done, 50th! well done, my majors!" (Applause.) There, after having completely routed the enemy, our gallant major fell to rise again, but totally disabled by five different wounds. He, however, afterwards returned to the Peninsula in 1809, and, at the action of Coa had two horses shot under him; he was again severely wounded at Busaco, was present at Fuentes d'Onor, the second siege of Badajos, and in very many brilliant skirmishes. (Cheers.) He served in a floating expedition on the coast of the United States of North America, landing many times to very good purpose; he was present in the campaign of 1815 in France, and was at the storming of Cambray; commanded the force employed in Scinde, and on the 17th of February, 1843, with 2,800 British troops only, he attacked and defeated, after a desperate action of three hours' duration, 35,000 of the enemy at Meanee (cheers); four days afterwards Hyderabad surrendered to his arms (cheers); and on the 24th of March, with but 5,000 men, he again attacked and signally defeated 26,000 of the enemy at Dubha, near Hyderabad (cheers), thus completing the conquest of Scinde. (Applause.) We find him again, in 1845, with a small force of all arms taking the field against the mountain and desert tribes on the right banks of the Indus, to the north of Shikurpoor, following them into the almost impregnable natural fastnesses of the Trukkar Hills, 1,000 feet above the level of the surrounding country, and effecting the destruction of those robbers; and, if permitted, with his little Bombay army assembled the following year at or about Bhawalpoor, he would have planted the British colours on the fortress of Moultan. Subsequently he was to be found when that dreadful scourge (cholera) attacked the troops at Kushawur, and when the 86th Regiment, in particular, had lost from the 15th to the 30th of June, 14 sergeants, 12 corporals, 1 drummer, 211 privates, 15 women, and 11 children, in all 264 souls. In the midst of disease and death, traversing the hospitals, seeing how best he could not only serve the poor sufferers himself, but, by example, how he could induce others to do so. These, gentlemen, are the acts by which a general is endeared to his followers. Sir Charles has been serving his country everywhere. His country owes him much; the soldiers not a little; for he has ever been the soldier's sincere and unflinching friend, and he is even at this time endeavouring to secure greater comforts for those who serve in India. It was my good fortune to have been sometime under his command in Scinde, and I can, therefore, say how much his troops admired him, native as well as European: they knew that he could not only take into and command them well in action, but lead them on to glorious victory. Gentlemen who have never been in India probably do not know that the gentleman in black (they always paint him white) is all-powerful; that, in fact, there is nothing impossible for him to accomplish. They had also a very exalted opinion of Sir Charles Napier's capabilities; but, as they could not exactly make him the gentleman in white, they would make him a very near relation, and, therefore, called him "Shytan kœ Bhae"—the Hindostan name for his satanic majesty's brother. (Laughter.) So you may all imagine, his name, with us, would have been equal to double our number in our projected attack on Moultan. Gentlemen, as I told you before, nothing in my power to advance can add to Sir Charles Napier's fame; still, I can only hope—and I am sure all my brother officers will join me in that hope—that if any enemy should be found rash enough to assail us, "the Conqueror of Scinde" may be selected as our leader, when, no doubt, he will, with his usual alacrity, again buckle on his trusty sword—perhaps that sword which was presented to him by Lord Ellenborough, while yet Governor-General of India, as a trifling acknowledgment for his services in Scinde—and provide himself with a good portable marching kit, not forgetting "the two towels and bit of soap." Then, gentlemen, I venture to say he will not forget his old corps, the 50th, but teach the young 50th what he taught the old 50 years ago—to fight to the death for their beloved Queen, their country, and her people—the finest country and noblest people in the world. (Cheers.) I now propose, gentlemen, "The health of our illustrious guest,

General Sir Charles Napier, the Hero of Scinde, with all the honours; may health, happiness, long life, and increasing honours attend him; or, to use an eastern expression, May his shadow never grow less." (Loud cheers.)

"The toast was drunk amid terrific applause.

"Band: 'See, the Conquering Hero comes.'

"Sir Charles Napier, on rising, was loudly applauded. He observed that he had to thank them all, in the first instance, for the way they had received the toast that had been given; and, he had to thank his friend, Colonel Sidley, in the second instance, for having spoken so favourably of him. He was afraid that all generals had a great many people under them who were partial to them, and he was afraid his friends there were. ('No, no,' and a voice, 'Right good reason for being so.') He was there among his family, he might say. (Cheers.) It was exactly forty-one years since he left the mess of the 50th Regiment, on the banks of the Tagus, in Portugal. The soldiers of that regiment were there drawn out to meet Marshal Soult, who, however, did not like their looks and went back. (Loud applause.) He (the gallant general) was very sick, not being able to ride back to his tent, and he was sent away from his regiment by his dear friend Colonel Stewart. He went off to Lisbon very ill, thinking his promotion was settled; but, however, he found a lieutenant-colonelcy, which took him away from the 50th Regiment, to his very great sorrow. He never rested till he got back to the regiment, a few years afterwards, but before he joined it—the 2nd Battalion—it was broken; and, from that, he had always looked with affection to the 50th and belonged to it in heart and feeling, and wherever he had been he had wished himself at the head, once more, of the 50th. (Loud cheers.) But, as events progressed, he certainly got a very awkward acquaintance with the 22nd Regiment, and he was in action with a large force. Those gentlemen who had been in India knew well how the Sepoys were run down. He had never seen them in action prior to the battle, he had heard much against them and very little in their favour, he could not, therefore, tell what their character in fight was. In the battle of Meanee he had had 500 British soldiers, the rest were Sepoys. Now, though they might hardly conceive it, the 500 men of the 22nd, who won the battle of Meanee by their courage, set an example to the Sepoys of unflinching daring, were followed by the Sepoys most gloriously, and, after the battle, he began to share his love for the 50th with the 22nd. (Cheers.) Well, the news of the battle had hardly reached England when the old general—Finch by name—gave up the ghost. He remembered that old general as colonel of the 22nd Regiment when he was a boy. The major-generals who had not a regiment thought that he would never die (laughter). Well, in the meantime, the last trumpet did sound, and the regiment being vacant Her Majesty graciously gave the 22nd Regiment to him; and in the interim they had fought another battle entirely with the Sepoys, except the remnant of the 500 British troops, whom, as they would suppose, were, by that time, tolerably diminished. And, in the second action the Sepoys were, as all knew who had been in India, the stand-fast of the battle. (Cheers.) Well, just after that he received the news that he was colonel of the 22nd Regiment; and from that time forth there was this difference between the 22nd and 50th Regiments in relation to him: the 50th were acquaintances in the abstract, it was the regiment he knew, but he never saw them after till he met them at Lahore. He loved the regiment, but he did not know the men or the officers; but the 22nd—men and officers—had fought by his side and won all sorts of honours for him in two desperate actions, and, therefore, he was obliged to tell them the truth, that the 22nd Regiment was the regiment he loved best in the world. (Applause.) That was to say, he loved the men and officers of the 22nd Regiment (applause), but his affection for the 22nd Regiment could never surpass the pride he had in the 50th Regiment. (Cheers.) He had great affection for both, and if the 22nd Regiment were not his own regiment he could hardly say which he loved the best. And now, having talked a little about Europeans, he thought it right to say, now that he was here, away from India and that he had done with India, that finer soldiers, braver soldiers, more well-conducted soldiers than the Sepoys of India he never commanded. (Loud applause.) He had commanded British regiments and seen plenty of them in action, and if it came to bayonets nothing could stand before them (cheers), and as far as firing went, and as far as following the British soldiers went, as of keeping up with them and emulating them, the Sepoys were always up to the mark, when well led. He trusted to the 22nd at the battle of Meanee; but he had such confidence in the Sepoys, and knowing the 22nd, he ordered the worst regiment of Sepoys to lead the attack. But he trusted to the circumstance of being present himself with his staff, and he had confidence in the Sepoys, he had no doubt of their conduct. He was, however, in doubt as to the issue of the conflict; and he changed the order of battle, and attacked with the left instead of the right, in which were the 22nd and 12th Regiments, and in both those actions the Sepoys behaved gallantly, and perfectly becoming British regiments. He did say that, wherever the army in India was led on by British troops, interspersed with Sepoys, and the general order of battle was well arranged, the Sepoys would never be found to be deficient in courage. He was particularly obliged to remark that he saw tried the 3rd cavalry of the Scinde Horse, the 12th Infantry, and the 25th Infantry; and that he would go into action with these regiments on the morrow, without a single British soldier. The Sepoys would fight well if they were well led, he was convinced; he had seen them tried in the hardest way, and never found them to fail. He saw the Bombay army, the officers of which some writers have lately ridiculed—they had risen from the ranks, and were, therefore, very old before they became officers—two years ago, scale the mountains and rocks, gallantly headed by the officers, old as they were, under a heavy fire and shower of rocks, doing their duty as well as it was possible for men to do it ; and he therefore thought it was very hard to hear a British officer, calling himself a Bombay officer, writing against those officers, because they were not, perhaps, as young as himself. They were not, perhaps, as young as himself, but they were as brave; they were not as active, because it could not be expected that he could get up a parcel of rocks as quickly as his friend Sidley could—(hear, hear, and laughter)—and, when any man had to scale rocks under a heavy fire, he did not go up so very fast, but took care to be sure-footed. However, these Sepoy officers went up most gallantly, and the other men followed. He kept her Majesty's troops in reserve, in case of any accident ; but it was only just to say that the Sepoys were right good soldiers and gallant soldiers in the field. Having thus dilated to them a little upon his military history, and the martial character of Sepoys, there was one subject in his friend's speech which he could not help noticing—that was, the death of Sir John Moore. He could hardly speak of that great man's death, because he was ashamed to open his mouth about it, after the beautiful description given of it by his brother, in his late work and his former work. The death of that great man was one of the most glorious events that ever happened. They talked about Wolfe's death : Wolfe was struck by a shot, and carried to the rear of the 22nd Regiment, when there was a cry from the heights of Abraham, 'They run, they run!' 'Who run?' asked the dying hero. 'The enemy,' was the reply. 'Then,' said Wolfe, 'I die content.' That was great and glorious. Well, when Moore was shot (he had this from Lord Hardinge, his aide-de-camp), his left shoulder was struck with a cannon shot, his left arm was hanging from his shoulder, and his heart was laid bare, and his agony was beyond description. He fell, sitting, and there he sat, looking upon the action, never asking for help, never uttering a groan. Hardinge dismounted his horse and ran to him, and said, 'I hope, general, this won't be mortal.' Moore looked round, and said, 'Hardinge, I am a dead man; tell me how the enemy get on?' He fixed his eyes upon their advance—because the 50th were always pretty forward—and made no other observation save 'Tell Baird that I am mortally wounded.' He never asked to be taken from the field. Lord Hardinge was going to take away his sword, which was dangling between his legs. Moore said, 'No, Hardinge, let it go from the field with me.' That was the only observation Moore made about himself. He was carried off the field by four men, in a blanket, and he made them turn round several times in sight of the scene of conflict, to let him look upon it. At that time his heart was laid bare, his arm was hanging by a piece of skin, and turned round the right shoulder. It was impossible to conceive a heroism greater than this. It was not from numbness produced by the shot, because, when he was conveyed into the room, the surgeon did not attempt to dress him, as the case was hopeless, and the dying general said, 'I feel so strong—I feel that I shall be too long in dying; I hope to God, it may come soon, to relieve me from this horrid torture.' So it was not numbness from the shot, but regular, right-down John Bull heroism. (Cheers.) It was some satisfaction to the young officers present, to hear one who was on the spot at the time speak of what happened to that great man. (Cheers.) His memory never could be too greatly held in estimation by the British army, particularly the 50th Regiment, to whom he gave the last words of encouragement in that great battle. (Cheers.) He thought he had then talked long enough. (Loud and long-continued applause.)"

Sir Charles Napier afterwards inspected the 50th at Fulwood Barracks, when he made the following address to them:—

"Soldiers of the 50th,—It would be useless in me to address

you as a steady and well-drilled regiment, as all regiments under his Grace the Duke of Wellington are well drilled. I address you as a hard-fighting regiment, as a regiment that has gained many a well-earned laurel in the field. It is now forty-three years since I commanded you as Lieutenant-Colonel and Major in many a hard-fought battle in the Peninsula. The other day I met you again at Lahore, and was exceeding glad to see that you still maintained the high renown, and general good disciplinary character, which you had previously so well earned on the Peninsula. No regiment in India—and I say it because you well deserve this tribute of praise—no regiment in India distinguished itself more than the 50th; and I have, on more than one occasion, had great cause to admire your bearing and action in the field, your patience and energy, your perseverance and moral courage. Soldiers, the regiment that a man first joins is his home, and he always afterwards looks upon it as his home; and when a man, after being absent from it for a length of time, comes back to it once more, his feelings are those of a man who is once more returning to the friends he loves—to the home which is endeared to his heart. I look at this visit of mine to this regiment as a return to my friends—to my old home; and, as being the greatest honour it is in my power to bestow upon this regiment, I now wear the same belt and plate I wore when major of this corps. [The general here exhibited to the soldiers the accoutrements mentioned by him.] I have again this day admired your steadiness and drill, and am truly proud to see you in the state you now are;—no one sick, no one absent, no one quarrelling, but rough and ready, able and willing to fight at the first sound of a gun."

He then turned to Lieutenant-Colonel Sidley, and complimented him in warm terms upon the state of the regiment, speaking particularly of their appearance and discipline.

Sir Charles then resumed:

"There has been a great deal of talk about the Minié rifle;

but I can assure you, 50th, there is nothing like 'Old Brown Bess,' with a fixed bayonet, a strong arm, a strong heart, and strong courage. I have seen the 50th, with the old ram-down musket and bayonet beat all before them, and they would beat the devil if they met with him. If you were to go to meet the French to-morrow I would be willing to lead you."

There is little more to relate. Sir Charles suffered from the wound he received at Busaco to the last day of his life.

He almost died of sunstroke shortly after his brilliant victory of Meanee; and as governor and commander-in-chief of Scinde, he was fighting the hill tribes at the same time as the 50th were fighting on the Sutlej; but he was never a favourite with the East India Company, and in 1847 he quitted his command, and retired to the seclusion of his family at Oaklands in Hampshire, until disaster in India again called him to the front.

In 1849 the Duke of Wellington offered him the post of commander-in-chief in India, and is reported to have said, "If you don't go, I must." He landed in India on the 6th May, 1849, and though the Sikh War was over before he arrived, he had much to do in the reform of administration, to which he lent all his energies. This made him many enemies.

He drew his sword for the last time in an expedition through the Kohat Pass, from which he returned successful, only to find a reprimand from the Indian Government awaiting him.

He at once resigned, and returning to England in March, 1851, he resumed his life at Oaklands, where he lived in retirement until he passed away on the 29th of August, 1853.

The annexed description of his end, from his life by Sir W. Butler, happily portrays the peaceful end of a gallant soldier:

"The full light of the summer morning was streaming into the room, lighting up the shields, swords, and standards of Eastern fight which hung upon the walls; the old colours of the 22nd, rent and torn by shot, moved gently in the air; and fresh with the perfume of the ripened summer, wife, children, brothers, servants, and two veteran soldiers who had stood behind him in battle, watched—some praying, some weeping, some immovable and fixed in their sorrow—the final dissolution; and just as the heroic spirit passed to Him who had sent it upon earth, filled with so many aspirations and generous sympathies, a brave man who stood near caught the flags of the 22nd Regiment from their resting-place, and waved those shattered emblems of battle above the dying soldier."

He was buried at Portsmouth on the 8th of September, 60,000 people attending his funeral (Sir W. Butler), and a monument in Trafalgar Square, principally erected by subscriptions from private soldiers, fitly commemorates the career of one who is reported to have said, "My pride and happiness through life has been that the soldiers loved me." (Sir W. Butler.)

APPENDIX II.

SIR HUDSON LOWE.

From "Napoleon at St. Helena and Sir H. Lowe," by W. Forsyth, M.A.

Sir Hudson Lowe belonged to a Lincolnshire family, but was born in Galway on the 28th July, 1769.

Shortly after his birth his father's regiment, the 50th, was ordered to the West Indies, and he was taken out with it.

On his return to England, before he attained his twelfth year, and while still at school, he was appointed to an ensigncy in the East Devon Militia.

In the autumn of 1787 he obtained a king's commission as an ensign in the 50th Regiment, which was then stationed at Gibraltar.

At this time writes Sir Hudson Lowe :—

"The fortress was still in a most ruinous state from the effects of the siege. The whole rock was literally covered with fragments of broken shells and shot, and there was not a house in the town nor a building within the batteries, which did not bear the marks of its devastation."

After having been more than four years on garrison duty, during which, he says, every third or fourth night was passed on guard, with no other appliance for repose between the relief of sentries than a blanket on boards and a pillow, generally resting on a stone, Lieutenant Lowe obtained leave of absence and travelled in France and Italy, whereby he obtained a knowledge of the language of these countries.

On his return to Gibraltar the war had broken out afresh,

and he proceeded with his regiment to Corsica, where he was actively engaged, until the 50th was ordered to garrison Ajaccio.

The future governor of St. Helena was thus quartered in the same town as the Buonaparte family.

Sir H. Lowe says, "We were all delighted with our change of quarters to Ajaccio. The town was well laid out, spacious, well built, and the citadel had excellent accommodation, though not sufficient for all the officers."

One of the best houses in the town was occupied by the mother and sisters of Buonaparte.

An officer of the 50th Regiment of the name of Ford, was for a short time quartered in their house, and spoke with much satisfaction of the kind manner in which the family acted towards him, the young girls, for such they were at the time, running slipshod about the house.

On the evacuation of Corsica, Lieutenant Lowe accompanied his regiment to Elba.

He obtained his company in 1795, and soon afterwards was appointed deputy judge-advocate.

When the 50th proceeded to Spain he was quartered for nearly two years at Fort St. Julien in Portugal. He was then ordered to Minorca, and was instrumental in raising the Corsican Rangers, to the command of which he was afterwards appointed with the temporary rank of major.

The Corsican Rangers under Major Lowe landed at Aboukir on the 8th March, on the right of the Guards. He was present with them at the battle of Alexandria, received the first proposals for the surrender of Cairo, commanded the rearguard of the escort to the French army on its march to Rosetta, and was present at the advance against and surrender of Alexandria.

At the peace of Amiens the Corsican Rangers were disbanded, and Major Lowe was placed on half-pay, from whence he was appointed to the 7th Royal Fusiliers. He was secretary of a Board at Malta from 1801 to 1802. He was next appointed on a secret mission to Portugal, to ascertain the military condition and resources of that country.

In September, 1803, he again proceeded to the Mediterranean to raise another corps of Corsican Rangers, of which he was appointed lieutenant-colonel on the 31st December, 1803. In this capacity he was employed in Sir John Craig's expedition.

He was afterwards placed with his corps in command of the island of Capri, from which place he was driven out by General Lamarque in October, 1807, after Murat had taken possession of his kingdom.

Colonel Lowe and the Corsican Rangers were then appointed to take part in an expedition against the Ionian Islands, and he was afterwards appointed governor of Cephalonia and Ithaca. He returned to England in 1813. The following year he was employed on missions to Sweden and Russia. He was afterwards attached to the allied Russian and Prussian army, under the command of Blucher until November, 1813.

Towards the close of that year he was ordered to Holland to organise Dutch levies, from whence he again joined Marshal Blucher on the 24th January, 1814, under whom he was present at thirteen actions, eleven being against Napoleon in person.

Colonel Lowe brought the news of Napoleon's abdication to England on the 9th April.

He obtained his brevet of major-general on the 4th June; and was appointed Quartermaster-General to the British troops in the Low Countries under the Prince of Orange.

He remained at this post in Belgium till the beginning of June, 1815. Assumed the command of the troops at Genoa on the 19th of that month. On the 2nd July Lord Exmouth's squadron appeared off that port. Sir Hudson Lowe sailed in the "Boyne" and landed at Toulon with his staff on the 11th July.

On the 1st August he received the information that he was to be entrusted with the custody of Napoleon Buonaparte on the island of Saint Helena, of which place he was appointed governor. The brevet rank of lieutenant-general and the command of the troops was given him, and his salary was fixed at £12,000 per annum.

I am indebted to "Napoleon at St. Helena and Sir H. Lowe," by W. Forsyth, M.A., published by John Murray, for most of the information in the above article.

APPENDIX III.

Early in December, 1848, General Viscount Hardinge, G.C.B., presented a splendid cup to the 50th (Queen's Own) Regiment, the gift of Prince Waldemar of Prussia, who was attached to the regiment through the greater part of the Sutlej campaign.

The cup was presented at a mess dinner at Dover Castle, and is thus described by the *Dover Telegraph* of that date.

The cup is of elegant form, about 24 inches in height, and weighs about 150 ounces. The foot, of silver gilt, is formed of the trunk of the palm tree, and upon a cluster of palm leaves, surrounding its upper part, is embedded the vase: this is of frosted silver, and upon its surface, exquisitely chased in *basso relievo*, are two scenes—the one of the battle of Sobraon, at the instant the 50th Regiment had charged the works, and succeeded in placing the regimental colours on their summit, and at the same instant that Sergeant-Major Cantwell, who bravely bore them, was shot, and fell in the arms of victory. The works also display the Sikh banner, which was afterwards captured, and adorned the room.

The second incident or scene is taken at the instant when, at the battle of Ferozeshah, the Sikh forces were yielding, and Doctor Hoffmeister, the medical attendant of Prince Waldemar, fell.

The cover is of burnished silver, and reposing upon the lid are the sleeping lions of the Punjaub, the whole being surmounted

GOLD CUP PRESENTED BY PRINCE WALDEMAR TO THE 50TH REGIMENT.

by a complete set of chieftains' chain armour,* modelled from that belonging to a captured Sirdar. The shield, the lance, the surcoat, and helmet, are all executed with the most exquisite correctness and delicacy. Round the margin of the vase is engraved: "This goblet is presented by His Royal Highness Prince Waldemar of Prussia, as a token of remembrance of the happy days spent amongst the officers of the 50th Regiment at Loodiana, and the following glorious campaign of the Sutlej."

After the usual loyal toasts had been honoured, Lord Hardinge presented the cup in the following terms, including a letter from the Prince. The reply of Lieutenant-Colonel Petit, commanding the regiment, and the further remarks of Lord Hardinge are also given, and will be read with deep interest.

Lord Hardinge rose and said: "Colonel Petit, having thus paid the honours which an assembly of Englishmen invariably with enthusiasm accord to Her Majesty the Queen and the rest of the Royal Family, it is now my pleasing task to present to Her Majesty's 50th Regiment of Foot the splendid cup before me. We all know the Prince, Prince Waldemar of Prussia, very well; in addition to being identified with us as a companion and a friend, the associate of our festive hours, he was associated with us in the glorious conflicts of Moodkee, Forozeshah, and Sobraon, and we were all witnesses of his distinguished bravery. When the Prussian doctor, Hoffmeister, his medical attendant, fell, he alighted from his horse and supported his body, pressed his hand, amidst a fierce fire, and, upon finding life extinct, he returned to the conflict. In fact, our acquaintance with this noble personage, only elicits his resemblance in bravery, and generous feeling to his great ancestor, Frederic the Great. I will take the liberty of reading the letter which has accompanied

* Very little of this chain armour now exists, the delicate and exquisite workmanship having worn away in the course of time. The description of the incidents depicted on the cup is very interesting, especially the fact of their being from drawings by the Prince.

the cup, in order that you may have the satisfaction of knowing his Royal Highness's sentiments on the subject:—

"My Lord, as you are a soldier, you must know that the tie of friendship between fellow soldiers will last for ever. I shall keep, therefore, always the warmest interest to one of the regiments of the Sutlej army in particular, in considering myself as a kind of fellow soldier of the officers of Her Majesty's 50th Foot. It is impossible to forget the happy days which I have spent among those officers at Loodiana, who obliged me greatly by the kindness and hospitality they bestowed upon me. I consider the most obliging of all their attentions, that they had elected myself and the gentlemen of my suite to be honorary members of their mess.

"So your Lordship will understand, that after having received so many proofs of kindness from the officers of Her Majesty's 50th, I am anxious to express to them my thankful feelings and highest esteem, as this regiment must be regarded as one of the bravest which fought on the Sutlej.

"I beg, therefore, your Lordship to deliver the accompanying goblet, which I have ordered purposely to be decorated with some sceneries of the different battles after my own drawings, to the officers of Her Majesty's 50th. I believe it will be most gratifying to them, who deserve such a distinction, and give still a greater value to the goblet if it is presented under these circumstances."

Lord Hardinge then said: "The gallant 50th have truly earned this compliment. I can confidently testify that in the presence of the most formidable field works at Sobraon, which it was ever my lot to have witnessed, the army had received a check, the other regiments in action heard a cheer, and then the gallant 50th made their appearance, with shouts of 'Make way for Her Majesty's 50th!' and then, assisted by the brave 31st, carried the works and annihilated the army of the Sikhs. With this and many other facts in my remembrance, I can truly say that I entirely agree with his Royal Highness. As if nothing should be wanting to add value to the gift, his Royal Highness has decorated the cup with scenes of your conflicts, from drawings of his own execution. Thus I have endeavoured to execute the trust imposed upon me; and now, Sir, I propose that we shall first fill the cup with generous wine, and that then we shall drink the health of that excellent, generous, and brave Prince, Prince Waldemar of Prussia." (Great cheering.) The cup having been filled and passed round the table,

Colonel Petit rose and said: "Lord Hardinge having proposed and drank his Royal Highness Prince Waldemar's health, I propose that we receive it with all the honours."

Mr. Latham: "I beg leave on the part of Prince Waldemar to return thanks for the distinguished honour which you have paid him. The Prussian Consul-General, Mr. Hobler, whom I represent, would have afforded himself the high gratification of being present, had not serious indisposition prevented. Commemorative as this cup is of the glorious events of the Indian campaign, and associated as it is with the noble Lord, your illustrious guest, I have full confidence that these glorious events will be kept in honourable remembrance by the brave and gallant 50th."

Lieutenant-Colonel Petit then rose and addressed Lord Hardinge on the part of the officers of the 50th Regiment, and said: "The officers of the 50th accept with pride and affection this tribute, and we beg that your Lordship will convey to his Royal Highness this inadequate expression of our feelings. We accept it, and we will most conscientiously preserve it with sentiments of affection, because, when we look upon this token, we shall be reminded of one who joined the British Army, emulous of the character of the British soldier in the field, and who was alike distinguished for his personal bravery as for his courtesy and gentlemanly demeanour. We shall look upon this cup with pride, as having received it from the hands of our illustrious guest, the late Governor-General of India, and it will serve as a beacon to lead us on to future glory, should our country command our services. We shall also look upon this cup to remind us of the dangers which we have mutually encountered in the field, and it will serve to draw together the bonds of brotherhood, which have existed between the British and the Prussian soldier; neither shall we forget that the tear of sympathy is due to the brave—we shall not forget that Dr. Hoffmeister administered assistance to the wounded of our regiment, after the battle of Moodkee. Gentlemen, I call upon you to fill a bumper. Officers of the 50th, I am sure you anticipate the toast I am about to propose to you, which is that of the veteran and noble Lord with whose presence we are honoured this evening, and one of England's most distinguished generals, whose deeds of conquest in India are now matters of glorious and immortal history. Gentlemen, the 50th Regiment have an especial interest in the high reputation of Lord Hardinge. You, officers of the 50th, have witnessed his conduct in the field; and the British nation has been proud to acknowledge his distinguished merit, in bringing to a speedy termination the glorious campaign. We toast the health of our Governor-General, Lord Hardinge, our leader in the field, and now our distinguished guest."

Band—"See the conquering hero comes."

After the enthusiastic cheering elicited by this toast had subsided,

Lord Hardinge rose and said, "Officers of the 50th, I am truly grateful to you all for the manner in which you have received the mention of my name by your gallant Chairman, as well as the manner in which you have drunk my health. I can only say that for this, as well as for your conduct in the field, I owe you a debt of gratitude. The Army of the Sutlej nobly did its duty, and the native portion as nobly emulated your distinguished conduct." He then concisely narrated the services of the 50th Regiment in the Peninsula under their gallant commanders up to the melancholy battle of Corunna, and his association with them. He recollected, as he passed along the lines with Sir Charles Napier, that he pointed out the gallant 50th to that gallant officer,* who, overpowered by his feelings, could scarcely articulate his sense of their bravery and the affection he felt for them. "I remember also," said he, "that this gallant regiment paid me the last honours which I received in India; thus, not only from my sense of meritorious services, but from affection, I feel identified with the 50th Regiment. To you, truly, I can apply a similar motto to that which decorates one of our household troops, second to none in the British Army. I beg to propose Sir Harry Smith, who commanded you."

The health of Sir Harry Smith having been drunk with cheers, Lord Hardinge then proposed, "Lord Gough and the Indian Army." He then described the additional strength of the Indian Army as compared with the time when it was first placed under his (Lord Hardinge's) command. Then it was only 44,000, whereas he left it with 54,000 men, in a high state of discipline, 120 pieces of artillery, eighty of which were siege guns; and he felt confident that when his gallant friend, who had won more victories in China and India than any commander, advanced to Moultan, as traitors and insurgents are never very brave, we shall be able to peruse with delight the glorious results. If any regret exists at present in his mind, it is the absence of the gallant 50th. "I therefore propose to you to drink Lord Gough and the Indian Army."

* A correspondent writing to the author on this subject says, as Sir Charles saw the 50th Regiment at Lahore reduced to about 150 men, it positively brought tears into his eyes.

APPENDIX IV.

PRISONERS OF WAR IN RUSSIA.*

On the 22nd December, 1854, two officers, Captain Frampton and myself, with about 70 men, formed the guard of an unfinished parallel of attack against the wonderfully perfect defence of our foes. All the early part of the night an awful fire of round shot, shell, and grape had been hurtling over our heads, as we sheltered as best we could. A few men were on sentry in front, and a sergeant's party were stationed some way down the ravine which separated the left from the right attack.

About 2 a.m. my attention was particularly drawn to what I thought, from the sound of musketry and vivid flashes of fire, was a sortie in force upon the right attack, and Frampton and self thought it best to get the men to stand to their arms, and be prepared in case a similar attack should be made on our part of the position. The men, at least many of them, had been lying asleep until now.

I passed along the rear of the line and was getting the party on the alert. When this was nearly done the outlying sentries fired, and immediately afterwards came tumbling in over the parapet, and almost as they did so a dark line of men was dimly seen in the darkness, advancing over the crest of the rising ground, along which the trench was traced. Fire was opened on them and a momentary halt ensued, but when engaged in front a mass of the enemy appeared on our right rear, and between these and those in front we were doubled up. Many men were killed and disabled. Frampton and myself, after a hand-to-hand fight, were knocked down and taken prisoners, and borne off by parties of the enemy.

* By Lieutenant-Colonel M. A. Clarke, 50th Regiment.

The remainder of that night was spent in a small house, used apparently as an officers' guard-room, within the Russian lines. In the *melée* I had got rather badly hurt, as well as much bruised, by the butt ends of firelocks. One ball went through my forage cap, another grazed my temple, and a third found a billet in my head, and though at the time I hardly felt it, I was laid up in hospital for some months, and it was not extracted until about five years afterwards, when again on foreign service in Ceylon.

The day after our capture we were taken before a Russian general, Osten-Sachen, and one or two other superior officers, and then conducted across the harbour to a fort on the north side called Severnaia, and lodged in the quarters of a Captain Kotzebue, a naval officer, who was most kind to us. He, I think, commanded in that fort. There we remained for ten days, and were then sent inland, passing through Baktchiserai, a town on the road to Simpherōpol, which was our immediate destination. Baktchiserai was the Tartar capital of the Chersonese, before it was brought under the dominion of the Tsar. The military commander lived in the palace of the ancient Khans or rulers of the country. The palace was a handsome building. Two days' travel took us to Simpherōpol, which is about seventy versts from Sevastōpol.

We travelled in a springless covered cart, and as the roads were merely tracks across country, the jolting was very great, and by the time we reached Simpherōpol my wound was much inflamed, and giving me great pain. We arrived long after dark, were taken to the residence of the Governor, who had apparently an evening party, and then to the hospital for officers, where those English officers who had the misfortune to be captured were accommodated in a ward to themselves.

There were in all five of us and an Italian officer, Captain L'Andriani, of the cavalry, who was desperately wounded in the thigh when charging with the English cavalry at Balaclava. We met with much kindness and civility from both military men and civilians, especially from a Russian, a professor in the Academy, and his English wife.

Most of the English officers had been wounded, but all were able to proceed into the interior long before I was, to a Government town, Raizan.

Each nationality had a different town assigned. The English, Raizan, the French, Kalouga, and the Turks, Tambov. Non-commissioned officers and men had other places assigned, but I think in all some three hundred of all ranks would have covered the number of our prisoners, and some of these were seamen of transports that had been wrecked in the terrible gales, that occasionally swept the Black Sea.

Several American medical men were doing duty with the Russian troops. They had entered the service temporarily to study gun-shot wounds. One of these I succeeded in getting to attend me, and much preferred to the little German Jew, who was my previous doctor.

A Russian Jew, a soldier, was told off as our servant, and a sentry was usually posted at night in a small room outside the ward, and through which the entrance was. This sentry usually slept, but ignorant as we were of the language, and with thousands of men between us and our own people, escape was hopeless.

English officers were at first allowed to walk about and pay visits. This permission was afterwards withdrawn, but I never learned the real reason for it.

Two of the Grand Dukes, sons of the Emperor, passed through the town and visited the hospital. Gortchakoff did so also He was on his way to take over the command from Menchikoff.

I heard the booming of the guns distinctly at the attack on Eupatoria, and saw some of the wounded arrive, as also a funeral or two the result of the action.

On the 1st April, 1855, 2 French officers, 85 soldiers, prisoners of war, English, French, and Turks, left under an escort of infantry for the interior of Russia. For the officers country carts were provided if we preferred riding in them to walking, but as the men marched we could only go at a foot-pace. We usually made sixteen or seventeen versts a day, halting every second or third day, but the length of the march was ruled by the distance from village to village. We were always quartered for lodging in the peasants' houses, and if there was not room they had to turn out for us. We instituted a sort of mess, and got one of the French soldiers to cook; the first one turned out a great scamp and had to be changed.

One of the French officers, Monsieur Veroudart, acted as caterer, he drawing all our government allowance of money, and dividing it occasionally if there was any surplus. The food was usually good, poultry, eggs, and milk, plentiful and moderate in price. A goose cost sixpence.

We gradually made our way northwards, passing by Perikop and Alexandrief. Near the latter a Madame Kankrini treated us very hospitably. We passed the night at her house, and she sent us on to our destination for the following day in one of her carriages.

We reached Ekaterinoslav on the 3rd May, where we were first conducted to the prison, where a small filthy room was assigned to us, with one small bed, a table, and nothing else, for three of us.

Late at night permission to go to an hotel arrived, kept by a Swiss named Meurice, and by him we were most kindly received.

Nothing could exceed the attention of him and his wife, and we lived at his table. We made many acquaintances, receiving uniform civility from all.

Two other English officers passed through whilst I was here, taken subsequently to Myselfi, but they were travelling post.

Late on the afternoon of the 7th we again started, and in sixteen days reached the next Government town, Kharkoff. The order of our marches was much the same as before, the escort only being changed.

We passed the towns of Novo Moskofski, Constantinograd, and Voltei, and on our arrival at Kharkoff, the usual delay took place before the authorities seemed to know what to do, and as this delay always took place in the main street, the populace were given an ample opportunity of gratifying their curiosity. We were at last conducted to an hotel, the St. Petersburg, and remained a week there. Professor Braillard and his wife were most kind, and I lived almost daily with them.

Here we parted company with the soldiers, and the French officers and I were forwarded the remainder of the way to our respective destinations, by telegas with post horses. These were springless carts, generally driven at the utmost speed of the horses, of which there always two, and sometimes three or more, attached to each vehicle.

The centre horse is in the shafts, the others capering on either side and always driven abreast; a bell is suspended to a half-hoop over the shaft horse. So long as the chaussée continues, the fatigue and jolting is bearable, but when going across country, I have found the pain sometimes almost insupportable.

A bundle of straw is the usual seat, but the better class of Russians when travelling in this manner generally provide themselves with pillows. The bottom of the carriage is made in a half arch, and when two people sit alongside of each other it is the acme of discomfort. The satisfaction of much more rapid travel reconciled one to the much greater discomfort, and in that vast empire the highest dignitaries were often compelled to travel in this manner, no other being available. In two days we reached Koursk, 200 versts. The day following our arrival we went to the orderly-room of the garrison battalion, called before the colonel in command, we were taken by him into his drawing-room, and in the course of conversation he made use of very strong language, insulting in the extreme towards the Emperor of the French. The French officers immediately rose, saying they could not possibly sit there, to hear the head of their State abused in such a manner. We got away with difficulty, but had not gone far on the way to our hotel, when an orderly overtook us with an order to go back, which we were compelled to obey. We were then ushered into the presence of the colonel again, who broke into a violent tirade of abuse, and Monsieur Des-Ecot getting rather excited and gesticulating, the colonel chose to consider that he was threatening him, and called in some private soldiers who were flattening their noses against the glass panels of the doors, and watching from the outside what was going on. He made some species of inquiry from them if they saw him threatened, which of course they said they had, upon which he ordered Des-Ecot to be taken away, and locked up in a cell. Veroudart, the other French officer, and myself were consigned to the guard-room. In the cell where Des-Ecot was placed there were also some Russian soldiers. After about a couple of hours' detention, we were ordered to dine with him, and had to sit down to table or return to the guard-room. Shortly afterwards he sent for Des-Ecot, and told him to sit and eat his dinner, which he flatly refused to do. Eventually some sort of compromise was effected. The colonel was very drunk at dinner, and must have been so all day judging from his conduct. His family seemed rather nice people, and very sorry for the way he behaved. When we left after dinner we saw no more of our drunken friend.

I then continued my journey, passing through Orel, Kalouga (where I bade adieu to the Frenchmen), Tula (celebrated for large iron foundries), and reached Raizan, after seventy-three days' journeying from Simpheropol. Raizan was the Government town allotted to the English officers. The governor was very civil, and had places laid at his table daily, at which any of us that liked to go were welcome, and placed no restraint upon our moving freely amongst friends in the neighbourhood. The only stipulation he made was, if going to a greater distance than twenty versts from the city, that notice should be left at his office. Raizan was a well-built, pleasant place, with many pleasant families, amongst whom we visited freely.

Rumours were current, after I had been a month or so here, that an exchange of prisoners was to be effected, and orders came from St. Petersburg that several of the officers were to proceed to Moscow, to see the ancient capital of Muscovy, and thence to Odessa for exchange. Moscow was about eighty miles from Raizan. I think the Imperial Government defrayed the expenses of five of us there, and the two others were told if they liked to pay their own they also might go. Chadwick, of the 17th Lancers, and myself were the two juniors, and consequently were left to pay our own way; but as neither of us could, money not being very abundant, I lost the opportunity of seeing that most interesting city.

Two or three days after the party for Moscow left; Chadwick and I also left on our long journey of 1,500 versts to Odessa, travelling post, and never staying more than a night on the way. A subordinate Government official, called a Chenovnik, was sent with us to make the necessary arrangements for horses, &c. We left on the evening of the 15th August, passing through Tula, Orel, Koursk, to Kharkoff, 712 versts, which we reached on the 22nd. Omitting Kalouga, a large portion of the way was by the same route I had previously travelled up country.

The Chenovnik who had accompanied us thus far was a very good fellow, who had his own carriage, called in Russian a "tarantass," much more comfortable than the telega. The body of the carriage is placed on two long poles, is consequently springy, and is provided with a hood.

At Kharkoff, which we reached late in the afternoon, we found the Moscow party at dinner, and they left in advance next day.

From Kharkoff our route lay through Pultava, where Charles XII. of Sweden suffered defeat, and the distinct trace of northern blood shows that many of his soldiers must have remained there, and intermarried with the Russians.

We continued our journey, reaching Odessa on the 1st September, got on board an English gunboat, which took us from that portion of the Tsar's dominions, and after sundry transhipments we came back to the Crimea, where I served with my regiment until the final evacuation.

English officers prisoners at different times:

Colonel Kelly, 34th Regiment.
Lieutenant Byron, ,,
Captain Frampton, 50th Regiment.
Lieutenant Clarke, ,,
} Taken in Sorties.

Lieutenant Duff, 23rd Regiment, Outpost, Inkermann.

Lieutenant Clowes, 8th Hussars,
Lieutenant Chadwick, 17th Lancers,
} Balaclava.

Captain Montague, Royal Engineers.
Lieutenant James, ,,

Dr. Easton, R.N.
Warren, ,,
} Taken at Hango.

Two or three Commissariat officers, I forget the names, and some merchantman officers from wrecks, were also taken.

APPENDIX V.

The following extract from a work, entitled "The Conduct of Major-General Shirley in North America," published in 1758, is from the pen of William Alexander, called Lord Stirling.

Lieutentant-Colonel Mercer, commanding at Oswego,* having received repeated intelligence that the enemy had some place or camp about thirty miles east of Oswego, and that on August 6th there was a large encampment of French Indians about twelve miles distant, despatched an express boat with a letter to the commanding officer on the lake, telling him he intended to send next day 400 men in whale boats to visit the enemy, and desiring him to keep to eastward to cover the men in the boats, but instead of complying the brigantine returned the next day, and in endeavouring to enter the harbour was driven by a gale on rocky ground, where she lay eighteen hours. Monsieur Montcalm, the French general, having intelligence that the brigantine was stranded, and the other two vessels returned into the harbour, took the opportunity of transporting and landing his artillery and troops in boats within a mile and a half of Fort Ontario, which, as the French officers declared after Oswego was taken, he could not have done had our vessels been out to the eastward.

Their artillery consisted of 32 pieces of cannon, from 12 to 24 pounders, and several large brass mortars and hoziets, which could not have been transported by land on account of the swamps, drowned lands, and creeks on the way. Their forces

* Fort Oswego was garrisoned by the 50th Regiment. (See plan in Chapter I.)

consisted of about 1,800 regular troops, 2,500 Canadians, and 500 Indians.

As disadvantageous accounts have been given of the behaviour of the garrison of Oswego (which consisted of part of H.M.'s late 50th and 51st Regiments and the New Jersey Regiment of Irregulars) during the time of the siege, it is a point of justice to those troops to set their behaviour in a true light.

August 11th, 1756.—About noon the enemy began the attack on Fort Ontario with the fire of their musketry, which was returned with small arms and eight cannon from the fort, and shells from the other side of the river. The garrison on the west side of the river was this day employed in repairing the battery on the south side of the old fort. That night the enemy were employed in making their approaches to Fort Ontario (by breaking ground and entrenching) and in bringing up their cannon against it.

August 12th.—At daybreak this day a large number of battoes were discovered on the lake on their way to join the enemy's camp.* Two sloops were sent out with orders to get between the battoes and the camp, but before they came up the battoes had secured themselves under the fire of the French camp, whereupon the sloop returned. The garrison on the west side this day was employed as the day before, and in the evening a detachment was made of 100 men of the 50th and 126 of the New Jersey Regiments, under the command of Colonel Schuyler, to take possession of the fort on the hill † to the south of the old fort, and under the direction of the chief engineer they were to put it in the best state of defence they could, in which work they were employed all the following night.

August 13th.—The enemy continued their approaches to Fort Ontario, and notwithstanding the constant fire kept on them

* These battoes contained Montcalm's 2nd Division, bringing artillery and stores.

† Fort Rascal. (See plan at p. 7.)

from the fort and the loss of their chief engineer,* killed in the trenches, had a battery of cannon ready to open within sixty yards of it. About 12 Colonel Mercer sent orders to evacuate the fort, first destroying cannon, ammunition, and provisions, and at about 3 p.m. the garrison, consisting of about 370 men,† managed their retreat across the river without the loss of a man, and were ordered to join Colonel Schuyler's force on the fort on the hill, and employed all the following night in endeavouring to compleat (*sic*) the work of that fort. The brigantine off the rock and repaired, but unable to go out, the wind continued to blow direct into the harbour.

This night and the night before the enemy made several ineffectual attempts to surprise the advanced guard and sentries on the west side of the river. On the eastern side of the river the enemy were employed in bringing up cannon and raising a battery against the old fort,‡ the garrison keeping a constant fire of cannon and shells on them from thence and the works about the fort. The cannon which most annoyed the enemy were four pieces § which were reversed on the platform of an earthwork that surrounded the old fort, which was entirely enfiladed by the enemy's battery on the opposite shore. In this situation, without the least cover, the train, assisted by 50 of the 50th Regiment, behaved remarkably well.

August 14th.—At daybreak the garrison renewed the fire of their cannon on that part of the opposite shore, where they had the evening before observed the enemy at work raising a battery, and the enemy returned the fire from a battery of ten 12-pounders, and were preparing one of mortars and hoziets. About 9 a.m. about 2,500 of the enemy passed over from east

* One account says shot by an Indian in mistake for an Englishman.

† Belonging to the 51st Regiment.

‡ Fort Oswego.

§ Marked "a" in plan.

to west of the river in three columns, in order to attack the garrison on the west.

On this Colonel Mercer, not knowing their number, ordered Colonel Schuyler with 500 men to oppose them, which would have been done, had not Colonel Mercer been killed by a cannon ball a few minutes after. About 10 a.m. the enemy's battery of mortars was ready to play, all the garrison's plans of defence either enfiladed, or mined, by the constant fire of the enemy's cannon, 2,500 of their irregulars and Indians in their rear ready to storm on that side, and 1,750 of their regulars ready to land in their front, under the fire of the French cannon.

Colonel Littlehales,* who succeeded Colonel Mercer in the command, called a council of war, who were, with the engineer, unanimously of opinion that the works were no longer tenable, and that it was by no means prudent to risk a storm with such unequal numbers.†

On beating the chamade the fire ceased on both sides, but the French improved this opportunity to bring up some cannon,

* This was probably the same Colonel Littlehales who was second in command of the former 50th or 7th Marines, disbanded at the peace of Aix la Chapelle.

† Another account says: "The next morning the Canadians and Indians crossed the river Oswego by a ford three-quarters of a mile above Fort Oswego. They were unopposed, as the movement was not discovered, and shortly afterwards showing themselves at the edge of the force yelling and firing, though too far to do much execution, they served to discourage. Up to this time the defenders, who probably only numbered about 1,000, had behaved with spirit, but about the time the commander Colonel Mercer was cut in two by a cannon shot, and the garrison, sick and disheartened, discouraged by the howlings of the Indians and the screams and entreaties of the women, of whom the fort contained about 100, and knowing that the place was quite untenable, decided to surrender. This they did on the 14th, Montcalm promising that they should receive 'all the regard which the most courteous of nations could show.'"

and to advance the main body of their troops within musket shot of the garrison, and everything was prepared for a storm. Hereupon two officers were sent to the French general to know his terms. He replied :

"The English were an enemy he esteemed, that none but a brave nation would have thought of defending so weak a place so long against such a strong train of artillery and superior numbers, and that the garrison might expect whatever terms were consistent with the service of His Most Christian Majesty."

During the whole time of the siege the soldiers behaved with a remarkable resolution and intrepidity against the enemy, exerting themselves to the utmost in the defence of the place in every part of duty, and it was with great reluctance that they were persuaded by their officers to lay down their arms.

APPENDIX VI.

ARTICLES OF CAPITULATION OF CAIRO.

Translation.

The commissioners above named having met and conferred, after the exchange of their respective powers, have agreed upon the following articles:

The French forces of every description, and the auxiliary troops, under the command of the general of division Belliard, shall evacuate the city of Cairo, the citadel, the forts of Boulac, Gizah, and all that part of Egypt which they now occupy.

The French and auxiliary troops shall retire by land to Rosetta, proceeding by the left bank of the Nile, with their arms, baggage, field artillery, and ammunition, to be there embarked and conveyed to the French ports of the Mediterranean, with their arms, artillery, baggage, and effects, at the expense of the allied powers. The embarkation of the said French and auxiliary troops shall take place as soon as possible, but at the latest within fifteen days from the date of the ratification of the present convention. It is also agreed, that the said troops shall be conveyed to the French ports abovementioned by the most direct and expeditious route.

From the date of the signature and ratification of the present convention, hostilities shall cease on both sides. The fort of ———, and the gate of the Pyramids, of the town of Gizah, shall be delivered up to the allied army. The line of advanced posts of the army, respectively, shall be fixed by commissioners named for this purpose, and the most positive orders shall be given that these shall not be encroached upon, in order to avoid all disputes; and if any shall arise, they are to be determined in an amicable manner.

Twelve days after the ratification of the present convention, the city of Cairo, the citadel, the forts, and the town of Boulac, shall be evacuated by the French and auxiliary troops, who will retire to Ibrahim Bey, the isle of Rhoda, and its dependencies, the fort of Foucroy and Gizah, from whence they shall depart as soon as possible, and at the latest in five days, to proceed to the points of embarkation. The generals commanding the British and Ottoman armies consequently engage that means shall be furnished, at their charge, for conveying the French and auxiliary troops as soon as possible from Gizah.

The march and encampment of the French and auxiliary troops shall be regulated by the generals of the respective armies, or by the officers named by each party; but it is clearly understood that, according to this article, the days of march and of encampment shall be fixed by the generals of the combined armies, and consequently the said French and auxiliary troops shall be accompanied on their march by English and Turkish commissaries, instructed to furnish the necessary provisions during the continuance of their route.

The baggage, ammunition, and other articles, transported by water, shall be escorted by French detachments, and by armed boats belonging to the allied powers.

The French and auxiliary troops shall be subsisted from the period of their departure from Gizah to the time of their embarkation, conformably to the regulations of the French army: and from the day of their embarkation to that of their landing in France, agreeably to the naval regulations of England.

The military and naval commanders of the British and Turkish forces shall provide vessels for conveying to the French ports of the Mediterranean the French and auxiliary troops as well as all French and other persons employed in the service of the army. Everything relative to this point, as well as in regard to subsistence, shall be regulated by commissaries named for this purpose by the general of division Belliard, and by the naval and military commanders in chief of the allied forces, as soon as the present convention shall be ratified. These commissioners shall proceed to Rosetta or Aboukir, in order to make every necessary preparation for the embarkation.

The allied powers shall provide four vessels (or more if possible), fitted for the conveyance of horses, water-casks, and forage sufficient for the voyage.

The French and auxiliary troops will be provided by the allied powers with a sufficient convoy for their safe return to France. After the embarkation of the French troops, the allied powers pledge themselves that to the period of their arrival on the continent of the French Republic they shall not be molested; and on his part, the general of the division Belliard, and the troops under his command, engage that no act of hostility shall be by them committed, during the said period, against the fleet or territories of his Britannic Majesty, of the Sublime Porte, or of their allies.

All the administrations, the members of the commission of arts and sciences, and in short every person attached to the French army, shall enjoy the same advantages as the military. All the members of the said administration, and of the commission of arts and sciences, shall also carry with them not only all the papers relative to their mission, but also their private papers, as well as all other articles which have reference thereto.

All inhabitants of Egypt, of whatever nation they may be, who wish to follow the French troops, shall be at liberty so to do; nor shall their families, after their departure, be molested, or their goods confiscated.

No inhabitant of Egypt, of whatever religion, who may wish to follow the French troops, shall suffer either in person or property, on account of the connection he may have entered into with the French during their continuance in Egypt, provided he conforms to the laws of the country.

The sick who cannot bear removal shall be placed in an hospital, and attended by French medical and other attendants, until their recovery, when they shall be sent to France on the same conditions as the troops. The commanders of the allied armies engage to provide all the articles that may appear really necessary for this hospital; the advances to be made on this account shall be repaid by the French government.

At the period when the towns and forts mentioned in the present convention shall be delivered up, commissaries shall be named for receiving the ordnance, ammunition, magazines, papers, archives, plans, and other public effects which the French shall leave in possession of the allied powers.

A vessel shall be provided as soon as possible by the naval commanders of the allied powers, in order to convey to Toulon an officer and a commissioner, charged with the conveyance of the present convention to the French government.

Every difficulty or dispute that may arise respecting the execution of the present convention shall be determined in an amicable manner by commissioners named on each part.

Immediately after the ratification of the present convention, all the English or Ottoman prisoners at Cairo shall be set at liberty, and the commanders in chief of the allied powers shall, in like manner, release the French prisoners in their respective camps.

Officers of rank from the English army, from his Highness the Supreme Vizier, and from his Highness the Capitan Pacha, shall be exchanged for a like number of French officers of equal rank, to serve as hostages for the execution of the present treaty. As soon as the French troops shall be landed in the ports of France, the hostages shall be reciprocally released.

The present convention shall be carried and communicated by a French officer to General Menou, at Alexandria, and he shall be at liberty to accept of it for the French and auxiliary forces (both naval and military) which may be with him at the above-mentioned place, provided his acceptance of it shall be notified to the general commanding the English troops before Alexandria within ten days from the date of the communication being made.

The present convention shall be ratified by the commanders in chief of the respective armies within twenty-four hours after the signature thereof.

Signed in quadruplicate, at the place of conference between the two armies, the 17th of June, 1801, or of the siege of Saffar, 1216, or the 8th Messidor, 9th year of the French Republic.

(Signed) J. HOPE, Brigadier-General.
OSMAN BEY.
ISAAC BEY.
DONZELOT, General de Brigade.
TARAYRE, Chef de Brigade.

Approved and ratified the present convention at Cairo, the 9th Messidor, ninth year of the French Republic.

(Signed) BELLIARD, General de Division.

APPENDIX VII.

ORDER FOR RAISING 50TH AND 51ST REGIMENTS IN AMERICA.—GEORGE R.

WARRANT FOR APPOINTING JOHN CALCRAFT, ESQ, AGENT TO YE TWO REGIMENTS OF FOOT TO BE RAISED IN AMERICA UNTIL OUR AGENT OR AGENTS SHALL BE APPOINTED BY COLONEL WILLIAM SHIRLEY AND SIR WILLIAM PEPPERELL RESPECTIVELY.

WHEREAS we have thought fit to order two regiments of foot to be forthwith raised in America, under the command of Colonel William Shirley and Colonel Sir William Pepperell, for the service and defence of our colonys (*sic*) there; and it being absolutely necessary for the good of our service that some proper person should be appointed to receive the monies growing due from time to time for the pay of our said regiments, and to pay and answer the demands and bills of the respective officers until an agent or agents shall be appointed by the said Colonels William Shirley and Sir William Pepperell respectively. And we, reposing great trust and confidence in the integrity and honesty of John Calcraft, Esq., have thought fit hereby to nominate, authorise, and appoint him to receive all monies from you for the purposes aforesaid until an agent or agents shall be appointed. Our will and pleasure, therefore, is that out of such monies that are or shall come to your hands for the use of our land forces, you pay unto the said John Calcraft, Esq., all such sume (*sic*) or sumes of mony as are or shall grow due from time to time for the pay of our said two regiments, until an agent or agents shall be appointed as aforesaid. And for so doing this, with the acquittance or acquittances of the said John Calcraft from time to time, shall be as well to you as to the auditors of our imprests, and all others whom it may or shall concern, a sufficient warrant and discharge.

Given at our Court at Kensington this 7 day of October, 1754, in the twenty-eighth year of our reign.

By His Majesty's command,

Mr. Pitt. H. Fox.

Extract from a letter of 4th November, 1754, to Sir John St. Clair, D.Q.M.G. of H.M. forces in America:—

"Lieutenant-Colonel Ellison of Shirley's and Lieutenant-Colonel Mercer of Pepperell's," &c., "have orders to repair in the speediest manner they can to Boston and New York, in order to hasten the levy of those two regiments."

The following warrant was received by Lieutenant-Colonel Ellison, 50th Regiment, in November:—

You are to proceed at once, in His Majesty's ship of Ware—— to Virginia with your——adjutant, and on arriving there are to take the speediest method of going to Boston, in order to levy the regiment of which William Shirley, Esq., Governor of New England, is appointed colonel, and immediately on arriving there you are to acquaint your said colonel and put into his hands the names of the officers already appointed and the blank commissions which His Majesty has been pleased to allow him to fill up with such persons as he shall judge most proper for H.M. service in their respective ranks; you will likewise put into his hands the letter which His Majesty's Secretary of State shall have delivered to you in relation to the levy money, provision masters, and carriages, &c., of the troops, and you will endeavour to raise and discipline the said regiment in the best and speediest manner you can, in conjunction with your colonel.

You will correspond with Sir John St Clair's Department,

Quartermaster-General of the forces in North America until the arrival of Sir John Braddock in Virginia, by every opportunity that you can, acquainting him with your success in the new levy and with what intelligence you may receive.

You will take care to make out weekly returns of the effective officers and men of your regiment, and not only enter them in your regimental books (specifying dates of commissions and enrolments), but transmit copies thereof to the General, in order to his sending the same to this office.

You will use your best endeavours to keep up strict military discipline in your corps, obliging your officers to keep the men regular, cleanly, and in messes, as being absolutely necessary for the good of the service.

Given at the War Office this 4th day of November, 1754.

By H.M. command,

P. 339 W. O. Book. H. Fox.

ORDERS AND INSTRUCTIONS FOR FORMING THE MEN OF SHIRLEY AND PEPPERELL'S REGIMENTS, NOW IN ENGLAND, INTO BODIES NOT LESS THAN 100 MEN IN EACH.—GEORGE R.

Whereas the greatest part of our 50th Regiment of Foot, commanded by Major-General William Shirley, and of our 51st Regiment of Foot, commanded by Major-General Sir William Pepperell, was made prisoners of war by the French, upon taking the Fort of Oswego in North America, and transported from thence to Quebec in Canada, from whence part of our said regiment has been sent by the Governor of Canada to England, in order to be exchanged for an equal number of French officers and men, prisoners of war, and part of our said regiments (*sic*) has been taken on board a French ship, in which they were embarked for old France as prisoners of war, and a part of these regiments has arrived, and part is now prisoned in France, and whereas we have already thought fit to send our orders to Major-General the Earl of Loudoun, Commanding-in-Chief our forces in North America, to incorporate such part of our non-commissioned officers and private men of our said 50th and 51st Regiments as were not in Oswego when it was taken by the French into our other regiments of foot now in North America, and his Lordship was to acquaint the officers that they were to go on half-pay, and his Lordship was also directed to place them to other regiments as vacancies should happen, at least such of them as should be thought fit for future service, and whereas our further intention is, that the non-commissioned officers and private men of our said 50th and 51st Regiments, now in England, with a certain proportion of the commissioned officers, be in like manner incorporated and annexed to such regiments of foot as we may think proper to send to North America. Our will and pleasure therefore is, and we do hereby direct, that you do repair to Totnes, and there, or in any other convenient place, assemble all such officers and men of our said 50th and 51st Regiments as are now in England, and make a review and take an exact muster of them, and send a state thereof to our Secretary of State at War for our information. Our further will and pleasure is, that you form all the said non-commissioned officers and private men of our said two regiments into bodies of equal numbers (not less than 100 private men in each), to each of which bodies you will appoint three lieutenants and an ensign, according to the list which accompanies this our warrant, and you will direct that these officers and men so appointed and formed be quartered at Totnes, and places adjacent, till such time as our future will and pleasure be known; and we do hereby direct that you sign discharges for all such non-commissioned officers and men of our said 50th and 51st Regiments as shall have a claim to the benefit of our Royal Hospital near Chelsea, and give them passes to come to

London to be examined, and that you likewise direct that there be given to each non-commissioned officer and private man, so recommended to Chelsea Hospital, one month's subsistance to carry him to London, where directions will be given for him being subsisted till he or they can be admitted to the benefit of the said hospital. Our further will and pleasure is, and we do hereby direct, that as soon as you have formed the said bodies or drafts you do acquaint the officers that we are pleased to order the establishments for the said 50th and 51st Regiments of foot shall cease from the 25th December, 1756, inclusive, and that all the officers of our said two regiments shall be entitled to and be placed on the British establishment of half-pay, from and after the 25th December, 1756, notwithstanding which we are graciously pleased to allow full pay to those officers to the day of your declaring this our will and pleasure as likewise a continuance of full pay to those officers who shall be appointed to the bodies of drafts before mentioned until such time as the said drafts shall be incorporated into other regiments. And you are to transmit to our Secretary of State at War a perfect return of the commissioned officers, non-commissioned officers, and soldiers you shall have so drafted and formed into bodies according to the above order, as also a list of the discharged men and of the officers who are to be placed on the half-pay. Our further will and pleasure is, that as soon as the rest of the officers and men of the before mentioned 50th and 51st Regiments, who are now prisoners in France, or in Canada, shall be returned or exchanged, they shall respectively be placed on the half-pay or be drafted into our other corps in like manner as the officers and men before mentioned have been.

Given at our Court at St. James this 25th day of January, 1757, in the 30th year of our reign.

By His Majesty's command,

BARRINGTON.

"LONDON, 13th August, 1782.

"His Majesty having been pleased to order that the 50th Regiment of Foot, which you command, should take the county name of the 50th or West Kent Regiment, and be looked upon as attached to that division of the county, I am to acquaint you that it is His Majesty's further pleasure, that you should in all things conform to that idea, and endeavour by all means in your power to cultivate and improve that connection, so as to create a mutual attachment between the county and the regiment, which may at all times be useful towards recruiting the regiment.

"But as the completion of the several regiments, now generally so deficient in this crisis is of the most important national concern, you will on this occasion use the utmost possible exertions for that purpose, by prescribing the greatest diligence to your officers and recruiting parties, and by every suitable attention to the gentlemen and considerable inhabitants; and, as nothing can so much conciliate them as a polite behaviour towards them, and an observance of the strictest discipline in all your quarters, you will give the most positive orders on that head. And you will immediately make such a disposition of your recruiting parties as may best answer that end.

"I have, &c.,

(Signed) "H. S. CONWAY.

"MAJOR-GEN. SIR T. SPENCER WILSON, Bart.,
&c., &c., &c.,
"Colonel of the 50th Regiment."

"HORSE GUARDS,
"25th September, 1827.

"SIR,

"I have the honour to acquaint you that His Majesty has been graciously pleased to approve of the 50th Regiment being hereafter called the 50th (or the Duke of Clarence's) Regiment, in place of the West Kent, in commemoration of the distinguished honour lately conferred upon the regiment by His Royal Highness in person, having presented the regiment with its colours.

"I have, &c.,

(Signed) "JOHN MACDONALD, D.A.G.

"Officer commanding 50th Regiment,
"The Duke of Clarence's Regiment,
"Portsmouth."

"HORSE GUARDS,
"22nd January, 1831.

"SIR,

"I have the honour to acquaint you, by direction of the General Commanding-in-Chief, that His Majesty has been pleased to command that the 50th Regiment shall be in future styled the 50th (or the Queen's Own) instead of the Duke of Clarence's Regiment, and that the facings of the regiment be accordingly changed from black to blue.

"I have the honour to be, Sir,
"Your most obedient, humble Servant,

(Signed) "JOHN MACDONALD,
"A.-General.

"To the Officer commanding the 50th (or
"the Queen's Own) Regiment,
"Templemore, Ireland."

APPENDIX VIII.

LIST AND SERVICES OF COLONELS OF THE 50TH REGIMENT.

Names.	Date of Appointment.	Remarks.
50TH OR 7TH MARINES.		
H. Cornwall . . .	11th April, 1741	Disbanded 27th October, 1748.
SHIRLEY'S 50TH, RAISED IN BOSTON.		
Shirley	4th November, 1754	Disbanded 25th December, 1756.
50TH (AFTERWARDS THE QUEEN'S OWN), RAISED AS 52ND REGIMENT.		
Studholme Hodgson . .	30th May, 1756 (Date of appointment to 52nd)	Major-General 25th June, 1759. Commander-in-Chief, capture of Belleisle.

APPENDIX VIII.—*continued.*

Names.	Date of Appointment.	Remarks.
John Griffin Griffin	24th October, 1759	Major-General 25th June, 1759. Commanded Regiment during Seven Years' War; afterwards commanded a brigade. Mentioned in despatches
Edward Carr	5th May, 1760	Major-General 21st February, 1757. Lieut.-General 22nd February, 1760. Commanded Regiment during Seven Years' War.
Sir William Boothby	5th September, 1764	Major-General 10th July, 1762.
Michael O'Brien Dilkes	3rd February, 1774	
Hon. George Moorsom	1st September, 1775	
Sir Thomas Spencer Wilson, Bart.	30th April, 1777	Major-General 29th August, 1777. Lieut.-General 20th November, 1782.
Sir James Duffe	31st August, 1798	Major-General 3rd October, 1794.
Sir George Townsend Walker, G.C.B.	23rd December, 1839	General 28th June, 1838. For services, *see* Lieut.-Colonels.

Sir H. Lowe, K.C.B., G.C.M.G.	17th November, 1842	Lieut.-General 22nd July, 1830. Commanded Corsican Rangers in Egypt, 1802. Governor of Napoleon Buonaparte at St. Helena.
Sir John Gardner, K.C.B.	20th January, 1844	Lieut.-General 23rd November, 1841.
Sir Dudley St. Leger Hill, K.C.B.	28th March, 1849	Major-General 23rd November, 1841.
W. Francis Bentinck Loftus	11th April, 1851	
James Allen, C.B.	11th October, 1852	
Right Hon. Sir George Arthur, Bart, K.C.B.	28th February, 1853	
Sir R. England, K.C.B., K.H.	20th September, 1854	Major-General 11th November, 1851. Commanded 3rd Division in Crimea.
Marcus J. Slade	12th November, 1862	Commanded 90th through Kaffir War of 1846-47. Governor of Guernsey.
Sir E. Walter F. Walker, K.C.B.	8th March, 1872	

LIST AND SERVICES OF LIEUTENANT-COLONELS OF THE 50TH REGIMENT.

Lieutenant-Colonels.	Date of Appointments.	Remarks.
John Littlehales	25th October, 1755	
John Mompeson	16th December, 1755	
William Wilkinson	19th April, 1758	
Richard Prescott	22nd May, 1761	
Thomas Calcraft	29th January, 1762	Captain at formation of the Regiment in 1756. Colonel 65th Regiment.
John Dalling	14th August, 1772	
John Gordon	19th February, 1776	
John Shee	12th June, 1782	
William Edmeston	13th June, 1783	
Patrick Wauchope	17th February, 1794	Two Lieutenant-Colonels. Lieutenant-Colonel Wauchope commanded the Regiment in Corsica at storming conventional redoubt, attack on Mozello Fort, and siege of Calvi. Killed at attack on Rosetta, Egypt.

John Rose	1st September, 1795	
Francis Erskine . . .	2nd January, 1796	Junior Lieutenant-Colonel.
George Townsend Walker . .	6th September, 1798	Vice Erskine. Transferred from 60th. Commanded 50th at Vimiero and Copenhagen, commanded a division in the Peninsula War. Severely wounded and G.C.B. at Badajoz.
Benjamin Rowe . . .	1st August, 1804	Third Lieutenant-Colonel. Commanded 2nd Battalion in its formation; Brevet Lieutenant-Colonel 1st January, 1801; commanded 1st Battalion in Egypt.
Charles Stewart . . .	17th February, 1805	Vice Rowe. Transferred from 2nd Battalion. Commanded 1st Battalion in the Peninsula till his death there; wounded at Vimiero.
Charles Hill . . .	13th June, 1811	Brevet Lieutenant-Colonel 25th July, 1810; commanded the regiment at Vittoria. Died in command at Jamaica after 41 years' service in it.
Charles Morland . . .	29th October, 1812	Vice Walker.
John Bacon Harrison . .	4th November 1819	Succeeded Lieutenant-Colonel Hill as the only Lieutenant-Colonel. Commanded the regiment in the Peninsula at the battles of Nive and Orthes.

APPENDIX VIII.—*continued.*

Lieutenant-Colonels.	Date of Appointments.	Remarks.
Nicolas Wodehouse	2nd September, 1824	Colonel 28th June, 1838.
Joseph Anderson, K.H.	1st April, 1841	Commanded Regiment at Moulmain. Commanded Brigade at Punniar. Severely wounded.
John Petit	19th September, 1848	Commanded Regiment at Punniar, Moodkee (horse shot), Ferozeshah, and Sobraon.
H. De Burgh Sidley	28th May, 1852	
Richard Waddy, C.B.	3rd March, 1854	Commanded 50th in Crimea—at Alma and Inkerman. Commanded Regiment in New Zealand. Afterwards Colonel on staff.
Henry Edwin Weare, C.B.	10th November, 1868	D.A.A.G. in Crimea. Wounded at Alma. Commanded 50th in New Zealand, including Rangiahwhia. Commanded a flying column.
Andrew C. Knox Lock	12th August, 1874	
John Thompson	21st June, 1879	
Arthur Evelyn Fyler	4th July, 1880	

DISTINGUISHED OFFICERS OF THE 50TH WHO DID NOT COMMAND.

Officers.	First Appointments in 50th.	Remarks.
Field-Marshal Sir Alured Clarke, G.C.B.	Ensign 20th March, 1759.	Commanded at the taking of the Cape of Good Hope. Commander-in-Chief in India. Colonel 7th Royal Fusiliers.
Lieut.-General Sir James Leith, G.C.B.	Captain 25th June, 1784.	Aide-de-camp to General O'Hara at Toulon. Held command in Spain during Peninsula. Governor and Commander-in-Chief of Leeward Islands. Colonel 4th West India Regiment. Died during command there.
Lieut.-General William Gifford	Ensign 9th April, 1789.	Aide-de-camp to General the Hon. Sir Charles Stuart in Corsica; afterwards in Portugal; and at the capture of Minorca. Commanded 43rd Light Infantry in Peninsula (under Sir J. Moore).

APPENDIX VIII.—*continued*

Officers.	First Appointments in 50th.	Remarks.
General Sir Charles Napier .	Major 6th November, 1806.	Commanded the 50th Regiment at Corunna. Severely wounded, and taken prisoner. Wounded on Lord Wellington's staff at Busaco. Commanded the force employed in Scinde, including the actions of Meance and Dubha. Colonel of the 22nd Regiment, and Governor-General of India in 1849.
Major Brevet Lieut.-Colonel Ryan .	. . .	Commanded the 50th in the Sutlej campaign after Lieut.-Colonel Anderson was wounded. Commanded the Brigade in which 50th were at Moodkee, Ferozeshah, and Sobraon. Commanded the Regiment at Alliwal. Severely wounded at Sobraon, and died from his wounds.

APPENDIX IX.

STATIONS OF 50TH (THE QUEEN'S OWN) REGIMENT.

Year.	Date of order to move.	Date of arrival.	Station.	Remarks.
1756	January 28th	Not known	Norwich	
	April 12th	April 24th	Dover	By march route
	October 12th	Not known	Harwich Languard Fort	By march route. Detachments to Huntingdon and Sudbury
	November 12th	Ditto	Ipswich	
1757	June 21st	July 6th	Chatham	By march. Encamped
	July 20th	—	Newport, Isle of Wight	Part by march. Encamped
		September 5th	Embarked on board fleet under Sir E. Hawke	
	October 15th (Disembarked at Portsmouth)	October 19th	3 comps. Guildford 3 ,, Farnham 3 ,, Dorking and Reigate	By march
1758	January 2nd (Marched)	January 7th	Maidstone	Ditto

APPENDIX IX.—*continued.*

Year.	Date of order to move.	Date of arrival.	Station.	Remarks.
1758	April 20th	April 21st	Headqrs. 5 comps. Canterbury 2 ,, Deal 1 ,, Sandwich 1 ,, Margate	Ditto
	,, 22nd	May 2nd	7 comps. to Dover	Ditto 2 comps. left at Deal
	September 28th	October 16th (Date of march)	Maidstone	
1759	June 16th	June 19th	Chatham	Encamped
	August 8th	—	Hdqrs. 6 comps. Canterbury 3 ,, Dover Castle	Comps. at Canterbury joined those at Dover on the 14th
	November 12th	—	4 comps. Farnham and Odiham 3 ,, Guildford 2 ,, Godalming	
	December 10th	December 13th	Hilsea Barracks	
1760	May 8th	June	Expedition in Germany under Prince Ferdinand	
1763	—	January	Cantonments on the frontiers of Holland	
		March	Landed in England and marched to Shrewsbury	

	May 3rd	—	Ireland	Embarked at Parkgate. During their stay in Ireland the regiment was quartered at Dublin, Athlone, Ballyshannon, Galway, Cork, and Kinsale, but dates are not given
1772	—	—	—	Embarked at Cork latter end of this year
1773	—	(Beginning of the year)	Jamaica	
			North America	Embarked at Port Royal
1776	May	—	—	Regiment broken up and drafted to other corps
	November	November	Salisbury (Staff of regiment only)	
1778	February 23rd	March 9th	Frome	
		May	Winchester	Encamped
		July 3rd	Embarked on broad H.M.S. "Centaur," "Vengeance," "Defiance," "Thunderer," and "Vigilance"	Embarked at Gosport. Apparently as marines
		July	Landed and marched to Exeter	
1778	—	August 7th	Topham	For assizes
		,, 17th	Exeter	
1779	—	March 12th	Tiverton	For assizes
		,, 23rd	Exeter	

APPENDIX IX.—*continued.*

Year.	Date of order to move.	Date of arrival.	Station.	Remarks.
1780	May 24th (Marched)	Beginning of June	Coxheath	Encamped
		November	Maidstone	
		April 28th	Sherbourne	
		June 2nd	Exeter	
		,, 25th (Date of march)	Brixham, Torbay	Encamped
		October 27th (Date of march)	Hdqrs. 2 comps. Totnes 2 ,, Modbury 2 ,, Pympton and Ridgeway 1 ,, Kingsbridge 1 ,, Brixham 1 ,, Newton Bushell	
1781	—	March 17th	Plymouth	In barracks
		June 6th	Roborough	Encampment near town
		October 29th	Plymouth	
1782	—	July 2nd	Raine (near Plymouth)	Encamped

		November 14th	Hdqrs. 4 comps. Launceston 3 ,, Lisguard 2 ,, Millbrook 1 ,, Saltash	
		December 31st	Plymouth	
1783	—	September 29th	Kinsale	Embarked in transports "Fame," "Whisk," and "Regard," on September 3rd; anchored September 22nd; landed 29th
	October 10th	October 10th	Charles Fort (Kinsale)	
	,, 27th (Began march)	December 2nd	Limerick	
1784	June 10th	June 13th	Kilkenny 4 comps. Clonmell 2 ,, Cashel 2 ,,	
	July 25th (Began march)	July 28th	Cork	
		August 21st	Gibraltar	Embarked at Cork on August 3rd under Colonel Edmondstone. Headquarters 3 comps. "Neptune"
1793	January (date of embarkation)	February 5th	St. Florenza, Corsica	
1796	—	October	Port Ferago, Elba	
1797	—	—	Gibraltar	
		June	Portugal	Quartered in Fort St. Julien for two years

APPENDIX IX.—*continued.*

Year.	Date of order to move.	Date of arrival.	Station.	Remarks.
1799	—	—	Minorca	
1801	—	March 8th	Aboukir (Egypt)	With the army under Sir Ralph Abercrombie
		October 17th	Malta	
1802	—	—	Belfast	Landed at Cork on May 4th, under Lieutenant-Colonel G. T. Walker
1803	November	—	Ballyshannon	
1804	June	—	Omagh	2nd battalion at Chichester
	(About) September	—	Enniskillen	
1805	July	—	Longford	
	August	—	The Curragh	
	September	—	Fermoy, Clonmel, Middleton, Fermoy, Mallow	1st battalion received large drafts from 2nd and marched to these places
1806	—	April	Cork	
1807	June	—	Deal	Embarked from Cork and landed at Ramsgate. All effectives from 2nd battalion drafted into 1st
	July 25th	August 16th	Copenhagen	
	—	November	Deal	

1808	December	—	On board ship	Secret expedition under Gen. Spencer scattered by gale. Some ships made Gibraltar and Sicily and some put back to England
	—	March	Gibraltar	
	May	—	Portugal and Spain	Embarked Gibraltar, touched at Cadiz, and landed at Figueras
1809	—	January	England	
	July	August	Walcheren	
	December	—	England	
1810	September	September 25th	Spain and France	Landed at Lisbon and joined army under Marquis of Wellington
1814	July 20	August 1st	Cork	Embarked at Poliac. France, disembarked in Cork Harbour. Establishment 800
	September 5th	—	Aughnacloy, county Tyrone	2nd battalion disbanded, remainder incorporated with 1st, December 25th
1815	April	—	Enniskillen	
1816	March	—	Londonderry	Establishment reduced to 600
1818	July	July 2nd, 3rd, 13th	Dublin	
1818	December 27th	(1819) March 7th	Jamaica	Embarked Cork January 7th and 8th, 1819
1819	,, 28th			
1827	January 12th	March 9th	Gosport	Depôt joined from Ireland
	,, 15th	,, 5th		
	February 15th	May 22nd		

APPENDIX IX.—*continued.*

Year.	Date of order to move.	Date of arrival.	Station.	Remarks.
1827	June 12th	—	Portsmouth	
	September 20th	September 29th	Weedon	New colours. Headquarters
	,, 21st	October 1st		
	,, 20th	,, 1st	Northampton	Three companies under Captain Gill
1828	July 24th	August 2nd	Manchester	
	August 3rd	,, 12th		
	July 24th	,, 1st	Stockport	Northampton detachment
1828	October 6th	—	Salford	Northampton detachment rejoined first
	October 15th	October 16th	Blackburn	4 comps.
	October 16th	,, 16th	Bolton-le-Moor	Headquarters and 6 comps.
1829	May 6th	—	Bury and Rochdale	2 comps. and returned 9th
	,,	May 12th	Accrington and Blackburn	
	,, 12th	—	Accrington and Haslington	2 comps. returned Bolton on 23rd
	,, 13th	—	Blackburn	Detachments were sent from Blackburn to Clitheroe, Preston, Oldham, Chorley, Bolton, and Ashton-under-Lyne
1830	June 28th	—	Liverpool	8 comps.
	,, 29th	—	,,	Headquarters and 2 comps.

	July 1st	July 2nd	Dublin	
	,, 3rd	—	Waterford	Headquarters and 6 comps.
	,,	—	Carrick-on-Suir	1 comp.
	,,	—	Felhard	1 ,,
	,,	—	Clonmel	2 ,,
	October 18th, 19th	—	Templemore	Headquarters and 6 comps., other detachments rejoined here
1831	March 16th	—	Ennis (co. Clare)	3 comps.
	April 1st	—	Newmarket, Fergus, Kilkishen, Quin, and Six Mile Bridge	Detachment which rejoined headquarters on the 30th June, 22nd and 23rd of July respectively
	May 6th	—	Silvermine	1 comp. rejoined headquarters 9th July
	October 3rd, 5th, 6th, 7th	October 6th, 8th, 10th, 11th	Athlone	2 comps. left at Thurles and Roscrea. Detachments were sent from here to Shannon Bridge, Roscommon, Ballinasloe, Casula, Curraghboy, and Gort House
	October 15th	October 18th	Dublin	3 comps.
	,, 21st	,, 25th	,,	Headquarters and 4 comps.
	,, 28th	November 2nd	,,	3 comps. Detachments sent from here to Newbridge, Maryborough, Carlow, Baltinglass, and Portarlington
1832	January 2nd	—	Naas	2 comps. and heavy baggage

APPENDIX IX.—*continued.*

Year.	Date of order to move.	Date of arrival.	Station.	Remarks.
1832	January 13th	—	Naas	Headquarters and 2 comps. Detachments rejoined here and were sent to Carlow, Castle Comer, Doonena, Bawn Yarrow, Newtown Barry, Mountrath, Maryborough, Athy
	June 5th	—	Dublin	All detachments rejoined here
	October 12th „ 15th	October 17th „ 19th	Birr	Detachments were sent from here to Tullamore, Banagher, Shannon Harbour, Meelick, Geashill, Mount Shannon, Loughrea, Portumna, Philipstown, Mullingar, Kinnity, Killemer, Ballinagar, Blue Ball, and Frankford
1833	May 13th „ 15th	May 18th „ 20th	Fermoy	Detachments rejoined here
	„ 30th, 31st		Ballinacurra (embarked at)	Sailed from Cove of Cork, 1st June

		June 6th	Chatham	A detachment sent to Harwich rejoined June 20th, 1834
1834	July 8th	November 21st	Windsor (N.S Wales)	Headquarters
1833 1834 1835	—	—	—	The regiment embarked at Chatham in detachments at different dates during these years for convict duty in New South Wales
1834	August 30th	—	New Zealand	2 comps. for special duty returned on the 11th and 13th November respectively
				Detachments were furnished from Windsor to the following places: Liverpool, Georges River, Sydney, Port Stephens, Wangello, Berrima, Bong Bong
				3 comps. to Launcestown, Tasmania
				2 comps. to Norfolk Island
1837	July 28th	July 29th	Sydney	Detachments rejoined here
1839	September 12th	—	Norfolk Island	163 rank and file and officers.
				This detachment was relieved by 96th Regiment at different periods early in 1840
1841	January 29th " 30th	May 6th April 17th	Chinsurah (India)	
1841	May 23rd	—	Fort William	

APPENDIX IX.—*continued.*

Year.	Date of order to move	Date of arrival.	Station.	Remarks.
	October 15th	November 12th	Moulmein	The party that sailed on the 15th went in sailing ships; the headquarters on the 19th in a steamer
	,, 19th	October 26th	,,	
1842	March 6th	March 31st	Fort William and Chinsurah	
	,, 10th	April 12th		
	July 19th	November 14th	Cawnpore	Moved by water up the Hoogley in nearly 100 boats
1843	November 14th	February 14th, 1844	On field service. Returned Cawnpore	Battle of Punniar occurred during this period
1844	October 15th	December 7th	Loodiana	
1845	December 14th	March 23rd, 1846	On field service. Returned Loodiana	During this period the battles of Moodkee, Ferozeshah, Aliwal, and Sobraon took place
1846	,, 18th	April 12th, 1847	Fort William	Previous to embarkation
1848	February	June 6th	Deal	Date of landing at Gravesend
		July 12th	,,	
	July 27th, 28th	—	Dover	2 comp. detachments at Hythe
1850	April 3rd	—	Portsmouth	Hythe detachment rejoined on 21st May
	,, 5th	—	,,	
	,, 8th	—	,,	

1851	January 23rd	—	Banbury	
	„ 27th „ 28th „ 29th	—	Athlone-under-Lyne	From this station detachments were sent to Stockport, Burnley, Burslem, Wolverhampton, Manchester, Bury. At Manchester a guard of honour was furnished to H.M. the Queen from the Stockport detachment on the 10th October
1852	February 24th „ 25th	—	Preston	
1853	June 13th	June 14th	Camp of Instruction, Cobham	Brigaded with 42nd and 95th, under Major-General Fane
	July 15th	—	Plymouth	
1854	January 14th	January 16th	Dublin	Here the regiment was ordered to recruit up to 1,000 rank and file
	„ 21st	„ 23rd	„	
	„ 28th	„ 31st	„	
	February 24th	March 7th	Malta	Headquarters and 6 comps.
	April 4th	April 11th	Gallipoli (Turkey)	3 comps.
	„ 7th	„ 12th	„	2 „
	March	„ 8th	Malta	2 additional comps. from Ireland
	April 9th	„ 15th	Gallipoli	Headquarters and remainder of the regiment

APPENDIX IX.—*continued.*

Year.	Date of order to move.	Date of arrival.	Station.	Remarks.
1854	„ 19th	—	Boulaher	Brigaded with 93rd and 2nd Battalion Rifle Brigade, forming lines of entrenchment
	May 6th	—	Camp Chifleck, near Gallipoli	Brigaded with Royals and 38th Regiment
	June 22nd	June 24th	Varna	By sea
	September 2nd	September 15th	Crimea, on field service	Battles of Alma and Inkermann took place during this period
1856	July 10th	August 11th	Portsmouth	Headquarters and right wing
	„ 12th	August 17th	Cork	Left wing
	August 21st	August 23rd	Kilkenny	
	November 5th	November 5th	Belfast	Headquarters and right wing
		„ 6th, 7th	„	Left wing
1857	March 27th	—	Dublin	
	July 1st	October 17th	2 comps. Colombo 2 „ Trincomalee	
	„ 2nd	„ 10th	3 „ Colombo	Headquarters and 6 comps.
			1 „ Point de Galle	
			1 „ Kandy	
			1 „ Newera Ellia	

1863	October 10th	November 14th	Auckland, New Zealand. On field service in New Zealand	New colours were presented to the regiment on board ship by Mrs. Waddy (wife of the colonel). Several engagements took place, especially Ranziawhia
1866	September 26th	October 14th	Brisbane	One company
	October 3rd	,, 9th	Sydney, New South Wales	Headquarters and 4 comps.
1867	June 3rd	June 10th	,, ,,	2 comps. from New Zealand
	July 11th	August 9th	Adelaide, South Australia	3 remaining comps. from New Zealand
1869	March 9th ,, 24th ,, 29th	June 14th	Devonport	The troops from Brisbane, Sydney, and Adelaide embarked at different dates on board the "Himalaya"
	December 9th	December 9th	Bristol	2 comps.
	,, 10th	,, 10th	,,	Headquarters and 3 comps. Detachment at Newport, Monmouth, and Trowbridge
	,, 10th	,, 10th	Exeter	5 comps.
1871	February 24th	—	Aldershot	2 comps. at Trowbridge
	March 1st	—	,,	5 comps. from Exeter
	,, 14th	—	,,	Headquarters and 1 comp.
			,,	2 comps. left at Bristol joined
1872	August 15th	—	Autumn manœuvres near Blandford	
	September 16th	—	Colchester	

APPENDIX IX.—*continued.*

Year.	Date of order to move.	Date of arrival.	Station.	Remarks.
1874	July 3rd	—	Aldershot	Depôt at Maidstone formed from this station on the 1st of August
	August 4th	August 8th	Dublin	
1875	July 29th	July 29th	Birr	1 comp.
	„ 30th	„ 30th	Curragh	2 comps.
	„ 31st	„ 31st	„	Headquarters and remainder of regiment
	September 16th	September 16th	Birr	Headquarters and regiment except 2 comps.
	„ 16th	„ 16th	Nenagh	2 comps.
1876	April 12th	—	Curragh	The regiment moved by companies from Birr to the Curragh as required for musketry, last company leaving on June 9th, non-effectives moving to Kinsale on June 8th
	August 18th	—	Kinsale	Headquarters and 4 comps.
			Camden Fort	2 comps.
			Bantry	1 comp.
		September 15th	Kinsale	1 comp. left at Curragh

1878	February 28th	March 4th	Edinburgh	Embarked from Queenstown, where detachments from Bantry and Camden Fort joined. A detachment was sent from Edinburgh to Dundee, which left that station on May 20th for Ballater as a guard of honour to Her Majesty
1880	March 19th	March 22nd	Colchester	*Viâ* Harwich. 1 comp. or detachment at Harwich rejoined headquarters on December 28th. 1 comp. or detachment at Purfleet

On the 1st of July, 1881, at Colchester, the title of the Regiment became

THE QUEEN'S OWN ROYAL WEST KENT,

and the number 50th was discontinued.

APPENDIX X.

CORSICA.

Despatch of Lieutenant-General Sir C. Stewart.

From *London Gazette,* September 2nd, 1794.

Calvi, *August* 10*th*, 1794.

Sir,

I have the satisfaction to inform you that the town of Calvi surrendered to H.M. forces on the 10th inst. after a siege of 51 days.

As I perfectly agreed with Lord Hood in opinion, that the utmost despatch was necessary in order to enable the troops selected for the siege of Calvi to begin their operations before the commencement of the unhealthy season, every effort was used to forward the necessary preparations, and so effectual were the exertions of the different departments, that in the course of a few days the regiments embarked at Bastia, and Captain Nelson, of H.M.S. "Agamemnon," consented in Lord Hood's absence to proceed to Agra, where a landing was effected on the 19th June, and in the course of the same day the army encamped in a strong position upon the Terra del Capucine, a ridge of mountains three miles from the town of Calvi.

From many of the outposts, and particularly from those the friendly Corsicans were ordered to occupy, I could distinctly discover that the town of Calvi was strong in point of situation, well fortified, and amply supplied with heavy artillery. The exterior defences, on which the enemy had bestowed considerable labour, consisted in the bomb-proof stone star fort Mozello, mounting 10 pieces of ordnance, with a battery of 6 guns on its right, flanked by a small entrenchment. In the rear of this line (which covered the town on the westward), on a rocky hill to the east, was placed a battery of 3 guns. Considerably advanced on the plain to the south-west, the Fort Mollinochesco, on a steep rock, commanded the communication between Calvi and the province of Balagin, supported by two frigates moored in the bay, for the purpose of raking the intermediate country. But the principal difficulties in approaching the enemy did not so much arise from the strength of the defences, as from the height of the mountain and rugged, rocky surface of the country it was necessary to penetrate; and so considerable were these objects against the usual mode of attack, that it was judged expedient to adopt rapid and forward movements, instead of regular approaches. In conformity to this plan of proceeding, the seamen and soldiers were laboriously employed in making roads, dragging guns to the top of the mountain, and collecting military stores for the purpose of erecting two mortar, and four separate gun batteries on the same night. One of these was intended against the Mollinochesco, the second to be constructed on the rocks to cover the principal one of 6 guns, which, by a sudden march and the exertions of the whole army, was to be erected within 750 yards of the Mozello.

From some mistake the battery against the Mollinochesco was built and opened two days before the appointed time, and considerably damaged that fort. Observing, however, that it was the determination of the enemy to repair and not to evacuate it, the Royal Irish Regiment was ordered on the evening of the 6th July to move towards their left, exposing the men to the fire of their artillery. This diversion was seconded at sunset and during the greater part of the night by a feigned attack of the Corsicans, which so effectually deceived the enemy, that they withdrew a considerable piquet from the spot where the principal battery was to be constructed, in order to support the Mollinochesco, and diverting the whole of their fire to that point, enabled the troops to complete their work. This important position established, the enemy was compelled to evacuate the Mollinochesco and to withdraw the shipping under the protection of the town. A very heavy fire immediately commenced on both sides and continued with little intermission until the 18th of that month, when observing that these batteries were considerably damaged, and a breach appearing practicable on the west side of the Mozello, a disposition was made for a general attack upon the outworks, under cover of two batteries ordered to be erected that night, which from their position would in the event of a check appear the principal object of the movement.

From the zeal of Lieutenant-Colonel Wauchope and the great exertions of the 50th Regiment, the battery which he undertook to construct within 300 yards of the Mozello, was completed an hour before daybreak without discovery; a signal gun was then fired from it for the troops to advance. Lieutenant Newhouse, of the Royal Artillery, with two field pieces, covered the approach; and the Grenadiers, Light Infantry, and the 2nd Battalion of the Royals, under the command of Lieutenant-Colonel Moore, of the 51st Regiment, and Major Brereton, of the 30th Regiment, proceeded with a cool, steady confidence and unloaded arms towards the enemy, forced their way through a smart fire of musketry, and regardless of live shells flung into the breach or the additional defence of pikes, stormed the Mozello; while Lieutenant-Colonel Wemyss with the Royal Irish Regiment, and two pieces of cannon under the direction of Lieutenant Lemoine, Royal Artillery, equally regardless of opposition, carried the enemy's battery on the left, and forced their trenches without firing a shot.

The possession of these very important posts, which the troops maintained under the heaviest fire of shell, shot, and grape, induced me to offer to consider such terms as the garrison of Calvi might be inclined to propose; but receiving an unfavourable answer the navy and army once more united their efforts, and in nine days batteries of 13 guns, 4 mortars, and 3 howitzers were completed within 600 yards of the town, and opened with so well-directed a fire that the enemy were unable to remain at their guns, and in eighteen hours sent proposals which terminated in a capitulation, and the expulsion of the French from Corsica.

It is with sincere regret that I have to mention the loss of Captain Sewcold, of the navy, who was killed by a cannon shot when actively employed in the batteries. The assistance and co-operation of Captain Nelson, the activity of Captain Hallowell, and the exertion of the navy have contributed greatly to the success of these movements. The spirit, zeal, and willingness with which this army has undergone the greatest labour and fatigue, in the most oppressive weather, is hardly to be described; and such has been the determined animation of both officers and men, that the smallest murmur has never been heard, unless illness deprived them from making their services useful to their country.

I am indebted to Lieutenant-Colonel Moore for his assistance on every occasion, and it is only a tribute to his worth to mention that he has distinguished himself on this expedition for his bravery, conduct, and military talent.

It is with the utmost confidence that I presume to recommend to H.M. my A.D.C., Captain Duncan, of the Royal Artillery, whose activity, zeal, and ability in his own and the engineers' department, merit the highest commendation and advancement.

Captain Stephens, the officers, and men of the Royal Artillery have distinguished themselves with their usual ability in the management of the batteries, and their attention to the different branches of that line.

Sir James Erskine and Major Oakes have been essentially useful in their different departments; and permit me to assure you that a cordiality subsists throughout the army which promises the most signal success on any future undertaking.

I have the happiness to inform you that Captains Macdonald

and Mackenzie and the other wounded officer and soldiers are in a fair way of recovery.

Captain Shand, an officer of great merit and my A.D.C., will have the honour of delivering this dispatch.

I have the honour to be, &c.,

C. STUART, Lieutenant-General.

Rt. Hon. HENRY DUNDAS,

&c., &c., &c.

From same Gazette.—Killed and wounded, 50th Regiment, 10th August, 1794 : 1 rank and file killed, 1 ditto wounded (previous to attack of Fort Mozello). At attack of Fort Mozello on morning of 18th July: 50th, 1 rank and file wounded. From 19th July to 10th August, 50th, none.

www.ingramcontent.com/pod-product-compliance
Ingram Content Group UK Ltd.
Pitfield, Milton Keynes, MK11 3LW, UK
UKHW021840270726
14058UKWH00002B/248